THE WASTE OF THE WORLD

Consumption, Economies and the Making of the Global Waste Problem

Nicky Gregson

First published in Great Britain in 2023 by

Bristol University Press
University of Bristol
1–9 Old Park Hill
Bristol
BS2 8BB
UK
t: +44 (0)117 374 6645
e: bup-info@bristol.ac.uk

Details of international sales and distribution partners are available at bristoluniversitypress.co.uk

© Bristol University Press 2023

British Library Cataloguing in Publication Data
A catalogue record for this book is available from the British Library

ISBN 978-1-5292-3243-1 hardcover
ISBN 978-1-5292-3244-8 paperback
ISBN 978-1-5292-3245-5 ePub
ISBN 978-1-5292-3246-2 ePdf

The right of Nicky Gregson to be identified as author of this work has been asserted by her in accordance with the Copyright, Designs and Patents Act 1988.

All rights reserved: no part of this publication may be reproduced, stored in a retrieval system, or transmitted in any form or by any means, electronic, mechanical, photocopying, recording or otherwise without the prior permission of Bristol University Press.

Every reasonable effort has been made to obtain permission to reproduce copyrighted material. If, however, anyone knows of an oversight, please contact the publisher.

The statements and opinions contained within this publication are solely those of the author and not of the University of Bristol or Bristol University Press. The University of Bristol and Bristol University Press disclaim responsibility for any injury to persons or property resulting from any material published in this publication.

Bristol University Press works to counter discrimination on grounds of gender, race, disability, age and sexuality.

Cover design: Andrew Corbett
Front cover image: Stocksy/Protonic Ltd

Contents

Acknowledgements		iv
Preface		vii
1	The Global Waste Problem and How to Think about It *Or, how to understand the 'too much waste' problem*	1
2	Discard, Social Order and Social Life *Or, discard is foundational to understanding waste*	31
3	Consumption, Consumer Practices and Consumer Discard *Or, how consumer discard relates to economies*	52
4	Conduits, Value Regimes and Valuation *Or, following consumers' discarded things*	76
5	Recommodifying Discard *Or, the challenges of turning discard into an economic good*	104
6	Waste, Money and Finance *Or, how turning discard into waste turns waste into an energy resource and an asset*	133
7	Future Directions *Or, rewiring waste through the three Ds (decarbonization, digital and discard)*	164
Notes		180
References		203
Index		227

Acknowledgements

This book has been in the pipeline a long time, not only because of the time it has taken to do the research that underpins it but also because it has required finding the time and space to write it. For that I have COVID-19 to thank. The pandemic gave me both the opportunity to leave academic employment and, through multiple lockdowns and restrictions, the time to devote to a book project of this type. In that I am acutely aware of my good fortune, of being someone for whom these times have been creative, immensely productive and transformative. Such experiences are the pandemic of the privileged. Many, including many of my former colleagues, have had a far harder time of it, while in the UK far too many people have died needlessly as a result of political choices. Yet, amidst the suffering and the loss, it is important to acknowledge experiences like mine. In the very early days of the pandemic, like many academics, I found research and writing hard to justify and difficult to concentrate on; it seemed a luxury and an irrelevance. But as the weeks became months and life became pandemic life, my thinking changed. It wasn't just that pandemic life and the writing life are made for one another. As the political attack on the social sciences, and the arts and humanities, has gathered pace in the UK, it has become more imperative to write – to show exactly why the social sciences matter and to do that while many of my colleagues have found it difficult, or nigh on impossible, to carve out space for that. 'Don't let a pandemic go to waste' has been a repetitive refrain of the last few years in UK higher education, mainly as shorthand to describe the actions of employers, who've used COVID-19 as the opportunity to restructure jobs and working conditions. I reappropriate it here as my motivation and inspiration.

In a book that draws on 30 years of research, it is impossible to do justice to everyone who has contributed to what that looks like in book form. To list you all would be a daunting task, no doubt marred by lapses of memory. In some instances, and I am thinking here of a litany of workplace visits, where close observation often said far more than what was being said by the escort, I don't even have a name to remember. Often there wasn't even a person that triggered the key insight. How does one thank something sprayed on a wall, for example? So I am not going to attempt this task. Rather, I offer my heartfelt appreciation to the individuals and organizations who participated in this research over three decades. Those conversations, observations and insights, be they spoken or not, have informed my journeying in the world of consumer discard, scrap, recovery and recycling, and waste in various parts of the world. Individually they offered fragments of this world; collectively, though, those fragments piece together. My hope is that this book demonstrates something of that whole.

More manageable are the debts of research as situated knowledge production. A great deal of the research that underpins this book was funded by multiple grants from the Economic and Social Research Council in the UK. Without that, very little of this work would have happened. The support funded a number of researchers over 15 years who accompanied me on the journey, each for a few years. None got to see the whole but we certainly lived the parts intensely – be that the summer of 'car booting' that was 1995, the great charity shop expedition of 1998, charity shop volunteering in the autumn and winter of 1998, the deep dive into retro fashion that was 1999, two years of what seemed like living with UK households and their stuff in 2003/4, and an equivalent period of time watching decommissioned naval ships being unmade in 2008/9. A massive thanks to Beth Longstaff, Kate Brooks, Alan Metcalfe, Helen Watkins and Melania Calestani for the work and for the fun and laughter along the way. Then, on *The Waste of the World*, a special thanks to Lucy Norris and to Meghna Gupta of Soul Rebel Films; Farid Ahamed, Raihana Ferdous and Nasreen Akter in Chittagong; Romain Garcier, Dan Swanton and Josh Reno; and not forgetting Kate Schofield and Zara Smith, who kept the show on the road. Then, and much more recently, a huge thanks to Pete Forman, for whom chasing down local authorities' municipal waste contracts became one of 2018's obsessions.

As academics we are nothing without the departments in which we work, and I have been fortunate to have done this research in two amazing but very different departments. This research has its origins in the Department of Geography at the University of Sheffield, where I worked from late 1990 to 2011. When I first set out on this journey, I've little doubt that most of my then colleagues thought I'd 'gone off on one' – who on earth wanted to know about car boot sales? At best, this was just some quirky pastime. But the beauty of this department is that it gave me the space and time to follow this trajectory and to take it in the direction it needed to go. Ever respectful and supportive, it was the epitome of a collegiate academic workplace. My thanks to my colleagues there and most especially to Doug Watts, Harvey Armstrong, Peter Jackson, Jessica Dubow and Matt Watson. In late 2011, I moved to the Department of Geography at Durham University. This pushed my work in new directions and opened my eyes to new possibilities. Numerous conversations with Gavin Bridge and Paul Langley have influenced my thinking more than they could ever know; so too, but in very different ways, have Alex Densmore, Ray Hudson, Rob Ferguson, Mike Bentley, Ben Anderson, Colin McFarlane, Louise Amoore and Cheryl McEwan. Then there have been the many hours spent talking all things 'impact' with Adam Holden, as we have tried – in between talking about music – to make sense not just of how knowledge connects with policy, but policy as practice and how critical social science can be put to work in this field. It's fair to say that without Adam finding a way to fund the contracts work, the evidence base for the financialization of municipal waste would not exist.

Beyond my immediate workplaces, a huge vote of thanks is due to those who supported my early consumption research, particularly Danny Miller at UCL and Kathryn Earle at Berg; and to those who saw the merits in thinking consumption through what I came to call ridding, particularly Alan Warde at the University of Manchester. Then there are my fellow waste travellers: Catherine Alexander, Zsuzsa Gille, Josh Lepawsky and David Evans. I hope our work has shown what critical social science has to offer waste studies. In the policy context, I want to thank the 'GoSci' team, particularly Mike Edbury. I will be forever grateful for what I learnt through the experience of working with you through 2016 and I hope I have put the lessons to work here.

In bringing the book to production, Rob Ferguson has proven yet again that he is the critical reader and proofreader every writer needs, three anonymous reviewers provided an excellent set of comments to fine-tune the manuscript, while the team at Bristol University Press, led by Emily Watt, has been excellent throughout.

Lastly, my biggest thanks and my greatest debts are to the two people who have been with me through the two parts of this journey: Louise Crewe and Mike Crang.

Preface

This book is a synthesis of 30 years' research. It's taken a while to write, not just because of the years involved but also because of what's at its core. Perhaps I was being a bit 'slow' but it took me a long time to realize that what I was working on was waste. Seemingly, I'm not alone. In 2020, I was interviewed by the editors of a handbook on waste scholarship in the social sciences, to help frame one of their introductory chapters.[1] One of their questions was, 'tell us how you got into researching waste'. Like the others they asked that same question to, my answers turned out to be very much of the type, 'I never really set out to look at waste'. I find that fascinating and instructive. It points to the elusive qualities of certain types of waste research. It also forewarns that interesting things start to happen to understandings of waste when the starting point isn't waste.

It is, of course, important to recognize that a great deal of academic research starts from waste. There is a huge body of scientific and technical work focusing on what is – or, I would argue, appears to be – self-evidently waste – material, physical stuff that needs to be either managed as waste or manipulated to become something that is not waste. Then there is the social scientific work that starts with waste. Although nothing like the size of the scientific-technical field, this is now a substantial literature. Mostly, it has looked at people in proximity to waste and the effects of that proximity. They are typically injustices and inequalities, be they the health inequalities that come from living and/or working alongside certain wastes or the socioeconomic causes and consequences of working with it. Another strand of social scientific research starts from waste as an environmental concern and addresses the governance and governing of waste. What unites these two strands of social scientific work is their passive positioning of waste. It's the stuff around which people gather and do all sorts of things: protesting about the siting of waste management facilities such as incinerators and landfills, campaigning about its illegal presence when it's been dumped, sifting through it to scavenge bits and pieces of value, making policy to shape what happens to it. Immediately, that flags up waste's elusiveness. While it's the apparent focus of study, really it isn't. What matters are the people and what they are doing, which means that waste, all too quickly, recedes from view.

My own work is not like this. Rather than beginning with waste, I stumbled into waste via research on consumption. This intellectual backstory is important because it functions as a necessary opening here, the academic positioning story for this book. It also frames my understanding of waste, the arguments that appear in this book and the return to consumption that features here. This may surprise some academic readers, particularly those who are familiar only with my work on waste. Let me start with that consumption work. My research then was part of a wider agenda that sought to move consumption research away from an overly narrow focus on retailing, marketing and branding, and to put social scientific thinking to work in enriching understanding of consumption. I began that work in the mid-1990s, researching an array of sites of secondhand exchange in the UK: car boot sales, charity shops (in other parts of the world these are known as thrift or goodwill stores), and retro fashion shops and arcades. Then, as consumption research became increasingly concerned with what people actually did with the things they acquired, I spent two years in the early 2000s working closely with households looking at how stuff falls out of people's homes. That project focused not just on the stuff that got discarded but on processes of devaluation and discarding, so why it was exiting the home and why it was going in the direction it did. Some of that stuff, of course, became waste, but a lot of it didn't, and what was extraordinary was that almost everyone who participated in that project – well over one hundred households – went to considerable lengths to make sure that a lot of this discarded stuff did not become waste. So, rather than chucking it in the bin or taking it to 'the tip', they took it to charity shops; they saved it up for a car boot sale or to pass on to a relative; they gave it away to a friend; or, this being the early days of digital platforms, they experimented with selling it on eBay. This research laid the groundwork for an approach that has been to see waste not as self-evident stuff but as a time-space contingent category of value and as stuff that becomes waste as a result of practices and processes of valuation.

While I was doing this consumption research, there were moments that pointed to a much bigger but barely visible world of discarded stuff and its circulation. These are the sorts of moments that any researcher encounters in projects. It's what doesn't fit the current frame, that which is in the interstices but has to be set to one side for later – if one is lucky. If I think back to these moments, three stand out, shaping the future trajectory of my work. First, there was the frequent reference in the charity shop where I volunteered through the autumn and winter of 1998/9 to Wastesaver – a place I didn't get to understand until much later, but that then was a mysterious black hole, into which went unsold clothing and donations that never made it to the shop floor. Then, second, there was a charity retail conference I attended in the very early 2000s, where I learnt from practitioners that only 20 per cent of donations were actually sold through UK charity retail outlets; the rest was absorbed by the international market. At the same time, and thirdly, anthropologists had begun to turn their attention to secondhand clothing, focusing on the street markets that are ubiquitous in many lower- and middle-income countries, where clothes that had been discarded in the global North became fashion in the global South.[2]

As I thought about those moments it became apparent that if discard was to be fully understood, then it had to be not only thought about in a different way, but also researched differently. To stick with the world of secondhand clothing, clearly there was a connection between a charity shop in Sheffield (and a charity retail chain in the UK) and the entrepreneurs selling secondhand clothing in Zambian street markets. Yet the two ends, in very different parts of the world, were all that the social scientific research of the time was seeing. To understand the processes by which clothes discarded in the global North became revalued as secondhand goods in the global South, one needed to focus on how they were connected: the international trade in discarded clothing. The world of international trade in secondhand goods, however, goes well beyond discarded clothing. It encompasses just about everything imaginable, including consumer goods and capital goods. It also has intricate connections with markets for the recovery of the physical materials from which these goods are made. What had become visible was the world of global resource recovery and recycling, or what has also been called the last great resource frontier of waste.[3] A veritable Pandora's box[4] had opened up. How do you research that?

Three fundamental principles have informed how I've gone about researching resource recovery and recycling. The first, and one that is far from unique to the field of waste studies, is to dispense with what is called 'methodological nationalism'. This is the default, and largely unexamined, approach that prevails across much of the social sciences and that, for pragmatic reasons, either limits research to what goes on within the boundary of one country or, at best, makes comparisons between a few countries. Sticking within the boundaries of one country, or even making comparisons between countries, is not much help when it comes to researching the international trade in secondhand goods, where stuff often moves in indirect, even disguised, ways. The second principle, and this is the direct counter to methodological nationalism, is to focus on the mobility of things and stuff – or, following things and the routes that things travel. Again, this approach is far from unique to waste studies. It underpins the entire mobilities canon of work in the social sciences, while there is also, for example, a huge body of social scientific research on commodity chains, in which this kind of 'following' work is fairly commonplace. That work traces the coming together of particular commodities as commodities that are sold in markets in the global North, from agricultural goods such as coffee, chocolate and tea, cut flowers, mushrooms and bananas, to assembled goods such as computers and mobile phones. The third principle, however, is unique to waste studies. This is that 'following' here involves the *dissolution* of things. Inevitably, that dissolution revolves around the materiality of things and what social scientists call their mutability – that seemingly self-evident things unravel to become multiple materials. A simple example: if I discarded the highly durable, outdoor-wear shirt that I'm currently wearing as I type, it might stay as a shirt, but equally it might not. Then it becomes textile fibres (100 per cent polyester, as it happens – no good for recycling), zips (nylon) and buttons, of an indeterminate polymer. At the other extreme of complexity is an example of a capital good: a container ship. When these things get discarded they result in a myriad of stuff, known and unknown: very large amounts of scrap metal (ferrous, copper wire, aluminium chiefly) but also asbestos, oils, plastic-sheathing surrounds to cables, marine engines and power systems, IT systems and communications systems, and everything that it takes for humans to live at sea – marine furniture, kitchens, cooking equipment, medical and dental equipment, interior wall boards, beds, carpet, flooring, toilets, showers and so on. Following dissolution, then, is a more complex activity than following the production of a commodity. It is also more geographically dispersed. Commodity production relies on standardized supply chains and production facilities, which, if not fixed, for capital investment reasons tend to hang around in particular places. Supply chains for discarded goods and scrap materials, however, are more transient, while facilities for repurposing or recycling can be geographically footloose. So, following things in their unravelling can sometimes feel as if it is taking one anywhere in the world, even though there are some predictable directions of travel.

The research project that set about investigating this Pandora's box provides the title for this book. The Waste of the World was a five-year programme (2006–11) funded by the UK Economic and Social Research Council, which focused on following things of apparent rubbish value and stuff that is categorized as waste. Along with colleagues, I've followed what we dubbed 'clapped-out ships' and container loads of discarded clothing destined for recycling markets in different parts of South Asia. We've been able to examine closely the conditions and circumstances of their destruction and revaluation as multiple goods, and pieced together the intricacies of the global recycling networks that connect the global North, China and the global South. I've also examined stuff that remains firmly in the waste category, for the very good reason that it is highly toxic to all life forms: radioactive waste and asbestos. Provided it's being managed appropriately, this sort of stuff does not move around much; rather, it tends to stay still because it is outside of economic circulation.

Travelling in this direction meant that my research shifted away from the cultural and social frameworks that inform much work on consumption, to engage much more with economic and political concerns. It should be said, however, that this work does not fit easily with the established concerns of economic research in the social sciences. Quite apart from the riposte from economists that waste is an externality and therefore beyond the economic domain, there is the challenge that such a focus brings to the many traditions of economic analysis in the wider social sciences. Researching global recycling networks leads to the economies of the global South, to networks of traders, to a myriad of enterprises based on recovery, refurbishment and repair, and to the intersections of this with domestic manufacture and consumption in the South. Not only does it challenge a model that continues to understand economy in terms of the templates provided by the economies of the global North, and which struggles to see micro entrepreneurship as meaningful economic activity, it also disrupts the main representation of the relationship between the economies of the global North and South in globalization, which is that the South acts as the manufacturing and assembly factory for consumption in the global North. Seen through global recycling networks, that model is inverted: consumption in the North supplies production (and consumption) in the South.[5] Then there is the challenge posed by storage. Mostly, economies are thought of in terms of flows of goods and money – of capital and labour, of exchange and transactions. What stays still is often thought of as if it is outside markets, as per the contents of a public museum or art gallery, or the things in our homes. In the real economy, though, storage isn't free. It's a cost – something that has to be minimized for financial reasons, hence 'just-in-time' supply chains. 'Forever-in-case' forms of storage, however, of the type required to manage certain toxic wastes safely, require storage in perpetuity. In these ways, we see how examining waste challenges and upturns many of the established frameworks that underpin knowledge in the social sciences. Yet, at the same time as doing interesting things to established fields, thinking through following discard does interesting things to waste studies. It acts back on the field, demanding reflection.

One of the characteristics of more humanities-inflected scholarship in waste studies is that some of it works through what is termed an 'ecological imaginary'.[6] Because it takes the materiality of waste seriously, this scholarship sees the world through a mesh or web of interconnection and flow. Unbounded and undisciplined, it produces work that transcends many of the established divisions of academic knowledge, insisting on their connection. Physical stuff (the domain of chemistry and physics, technology and engineering) is absorbed within and mediates animal, plant, marine and human life (the domains of biology, molecular science, health and the social sciences). This pattern of interconnected thought, however, mostly stops short when it encounters economy. So, while there has been work that has taken up the challenge of following the dissolution of things economically[7], mostly what we find in waste studies is the persistence of, or at least the trace of, the unidirectional trajectories and the distinctions that inform understanding of economy. There is an emphasis on the distinctive wastes generated by extractive industries and agriculture, by manufacturing industry and by the activities of consumption, which generate 'post-consumer' (or 'municipal') waste. There is the tendency to think of waste by sector, so construction waste, demolition waste, transport waste, food waste, and so on. And there is also an emphasis on 'waste by industry', from the chemical industry, from the steel industry, and so on – a tendency that has been reinforced by efforts to make producers responsible for the wastes they generate. In recognition of the volumes of waste that are generated by many extractive and productive activities, for many in the waste studies field, the emphasis that was placed early on in its development on post-consumer (or municipal) waste is now seen as misdirected.[8] Attention, therefore, has turned to the wastes of what, before the advent of financial capitalism, used to be called the productive economy. To be sure, much of the research that comprised The Waste of the World programme could be seen in that same way – as an argument for exactly this focus. The contention that informs this book, however, is that this risks throwing the metaphorical baby out with the bathwater. Stated more strongly, there is a need for waste studies to return to municipal waste and reclaim it, not by corralling it as municipal waste but by thinking about it through webs of interconnection and flow. There are at least three reasons for this.

First, there is the imperative to connect municipal waste with what's going on elsewhere in waste studies, not abandon it. It is not even necessary to think about economy in terms of webs, connection and flow to see why this needs to happen. The basic laws of economic supply and demand tell us that what is driving the volume of waste generated by extractive, agricultural and manufacturing activity is not extraction, agriculture or manufacture per se, but rather global demand – for resources, agricultural commodities, basic manufacturing commodities and, increasingly, as more people in the world consume more, consumer goods. Without this demand these wastes would not exist. It is the global demand for plastic, for example, and the ubiquity of its applications that results in complex and problematic polymers occurring in multiple waste streams across the world and/or manifesting as marine pollution. Likewise, increased future global demand for electric vehicles will see an inordinate increase in extractive activity related to the mining of nickel, copper, cobalt and lithium, and thus a parallel increase in mining waste[9] while simultaneously requiring more steel, aluminium, plastics and textiles for their manufacture. In turn, down the line, as manufacturers' requirements to replace vehicles kick in, those same vehicles will become post-consumer waste.[10] It surely matters that we connect that post-consumer waste to its cumulative legacy in waste. Waste studies finds itself in as good a place as any field to draw those connections, but only if it reconnects with municipal waste.

Second, there is a need to acknowledge municipal waste's connections to consumption, and to consumption as an economic activity. Rather than see municipal waste as *post*-consumer waste and as somehow outwith or beyond economy, in this book I position it very differently – at the heart of many economies. I see municipal waste not as an end point but as a dynamic effect of what I term 'consumption-heavy' economies. When we look at the economies of the global North – at the breakdown of employment, or jobs, and at the contribution of particular sectors to GDP – then we see how important consumption is to economic life in those countries. Economic growth has become dependent upon expanding and intensifying the consumption of goods and services; the majority of jobs and livelihoods depend upon it. It's increasingly the same story in the world's biggest economy, China, where the strapline of globalization, 'made in China', has been adumbrated recently to include 'and sold in China'. This matters because it will lead to more and more municipal waste. This is not just because more consumers consuming more goods spells more waste. Increasing the number of consumption-heavy economies will require consumers to generate more and more discard to maintain economic growth. This is the reason why this book sees me returning to my previous research on consumption.

The third and final reason why there is a need to return to municipal waste is that it is here that the effects of the rise of financial capitalism on waste can be seen most clearly. In my most recent research, I have come back to municipal waste (in England) to show how contemporary waste management solutions relate to finance capital. Much as is the case with other utilities, municipal waste offers huge opportunities for the finance sector. It should come as no surprise, then, to find that, much like household water, municipal waste has been financialized. The significance of this goes beyond the immediate point that financialization demands, at minimum, the continued generation of municipal waste, and preferably its expansion. It is this: that in turning away from municipal waste, the field of waste studies was missing its connection to money and finance, and to a global economy that now dances to the tune of finance capital. It is my contention that if waste studies is to advance its understanding of waste's relation to economy beyond seeing waste as a resource, then it needs to revisit municipal waste.

★ ★ ★

What follows, then, is a book that is mostly about what waste studies classifies as municipal waste. For those readers who may be familiar with my work on waste, the synthesis offered here therefore is not a résumé of all of it. Parts of the research outlined earlier, notably that on toxic waste, do not appear here except in passing. The reason for that is not just that the book is mostly about municipal waste; it's also organized around a central problem: the global waste problem, as this is understood by policy practitioners, that is, as growing mountains, or a tsunami, of municipal waste.

My reasons for taking this problem-centred approach need making explicit. Some are generic. Municipal waste attracts political attention everywhere in the world. This is because it's what people understand and what they can relate to; it's what they produce, it's tangible, and when systems of collection are disrupted, fail and/or are non-existent, it accumulates, affecting lives in profoundly negative ways and impacting quickly on health and well-being. As a result, it is this type of waste that politicians around the world find themselves having to pay attention to, while policymakers are faced with the challenge of finding solutions to the problem of too much (municipal) waste. But there are particular reasons for this choice too.

Through 2016, I had the privilege of working closely with the UK Government Office for Science, a body that works across UK government departments under the direction of the Government Chief Scientific Adviser, as part of a team responsible for preparing a report that sought to push the agenda of enhancing resource productivity by using wastes more efficiently.[11] That gave me insight into the workings of the UK policy world, and writing for it and within it, albeit in a pan-departmental context. The experience affected my understanding of social scientific knowledge and practice. It also challenged my understanding of the policy world and its relationship to the academy. Let me begin with the latter.

Prior to working within the policy environment, my understanding of the policy world, and of its implications for academic practice, was one that will be familiar to academic readers in the social sciences. My understanding was to see the responsibilities of academics as reporting and disseminating the *findings* from research (not the argument) and to seek to engage with policy practitioners to suggest specific interventions based on those findings. Research findings in this model translate seamlessly to policies. Effectively they seek to bypass the entire policy process and the conditions of its possibility, which are both political and economic. It is no small wonder, then, that most of those suggested interventions come to nothing, because academic research is often out of kilter with the policy cycle in that particular area or department; because of an absence of close ties between academics and policy practitioners, or its converse – close ties that grant privileged access to the policy world but that exclude other voices; because politicians don't see any votes in these suggestions; and because suggested interventions are based on research, which means that they are almost always uncosted, singular proposals, rather than a carefully costed set of alternatives, modelled on cost–benefit terms.

Working and writing within the policy environment places one within the policy process, albeit on the margins. It establishes the contours, or the interface, between the academy and policy practitioners; the meeting of, and distinctions between, these different types of expertise; and, above all, the important work of translation that is central to making evidence-based policy. If there are 'take-homes' from my involvement in this process, they are: 1) that research findings travel into policy via their translation into what are called 'key messages'; and 2) that policymakers, unsurprisingly, are confident in their ability to formulate policy, that's their expertise. Where they are less confident – again, unsurprisingly – is in their wider understanding of a problem, or what is called its 'framing'. This is where academics can help, by providing that kind of framing and by translating the findings from research into clear messages that policymakers can use to devise potential interventions.

Identifying messages is not an easy task for academics, who are accustomed to conveying arguments and/or findings. So, this needs illustration. Take as an example the COVID-19 pandemic. Early on in the pandemic, when little was known about the virus and its means of transmission, messages were conveyed that it might be being picked up from surfaces, hence the range of interventions that emphasized handwashing, not touching faces and the use of hand sanitizer in all public spaces. Now, with transmission understood to be airborne (a research finding), the message has switched to the need for policy interventions to target air and ventilation. The message, then, is the implication of a finding; it is not to be confused with specific interventions.

My experience of writing for and in the UK policy world also challenged my understanding of social science as knowledge and as practice. It destabilized the academy's internal narratives of what it does with an 'outside-in' view of the value of that knowledge and practice. This has implications; although, inevitably, implications will be context dependent. They can, and will, vary according to the character and status of social scientific research in different countries and the value placed on the academy by the policy domain. I am acutely aware, therefore, that what follows is UK specific and that my understanding is also shaped by having spent the entirety of my career employed by, and working in, universities in the UK.

One of the major challenges facing UK social sciences currently is the same as for every other area of academic activity: it's about the value placed on expertise. In a world that has become more strident, more opinionated, more tribal and altogether more instant and transient in its modes of communication as a result of social media, a cacophony of voices now compete with academic research and scholarship. But the challenge goes deeper. The raison d'être for much academic work is the production of knowledge. This is grounded in the rigorous application of appropriate methods pertaining to a recognized problem, or area of uncertainty, the analysis of data and an equally robust evaluation of evidence. In the face of a mounting populism across many parts of the world, respect for this kind of knowledge, for those who work in the academy and for experts can no longer be assumed. As a result, such knowledge is increasingly countered by, and in some quarters regarded as no more than equivalent to, opinion. This became abundantly clear in the UK in the response to the COVID-19 pandemic, in which science found itself under attack from a variety of perspectives.

Being heard in UK policy circles is not generally a problem for science, though being listened to most certainly is. However, for the social sciences, even being given the space to be heard can be difficult. For sure, some social science voices do get a seat at the table. These are mostly economists and psychologists. I don't think that is a coincidence, for three reasons. First, there is the type of knowledge these disciplines produce. These are the disciplines in the social sciences that most closely adhere to the model of scientific knowledge production. Second, they are the social science disciplines that have individuals at their core and that, if they make space for a sense of the social, see this primarily in terms of individuals' relationships to groups. Third, this focus accords with the political moment; it chimes with some of the central tenets of neoliberalism. So, findings find listening ears, even while listening is still highly selective. Hence, the behavioural model that has found favour in government circles in the UK is the least cost, most highly individualized, encoded in the idea of the behavioural 'nudge' and personal responsibility.

By contrast, the rest of us social scientists (sociologists, geographers, anthropologists) are mostly absent from the UK policy table. Typically, we have explained that absence away in terms of: 1) the 'critical' nature of the knowledge we collectively produce; and 2) the methods we use. There is some truth in the first of these in that the positionality of critique frequently sits uneasily with, and in tension to, the sites and spaces of power and the solutions focus that is favoured there. But the second is now more contentious. Whereas that might have once been the case in the UK – with qualitative research being relegated to the status of 'anecdata' – it is now far from a view universally held. The contribution of anthropology to understanding and mitigating the transmission of Ebola in West Africa is widely seen as a turning point.[12] Yet, more significant than either in explaining our absence, I would argue, is that too much of current social scientific output falls short with regard to the rigour of its evidence base, in the clarity of its findings and in moving findings to become wider messages, if it even bothers to identify those. Too often, being critical is seemingly enough; our analyses stop here. Too often, what passes as critical social science knowledge is actually better understood as commentary and opinion, not evidenced research. There is nothing wrong with commentary and opinion, of course, but it fails the evidence-based litmus test that underpins and differentiates commentary from knowledge. And then we tend to write in jargon. This speaks to initiates who know the language and who use it, but its meaning is opaque to the world beyond. That's all very well if the jargon actually has a meaning and can be explained and translated, but when jargon, and most especially neologisms, becomes the endgame (as has happened in parts of the social sciences) then we have a problem. This kind of stuff just doesn't travel well beyond the borders of its use because it doesn't produce easily comprehensible messages of the type that policymakers might understand and be able to translate into actions. If anything, all it does is reinforce prejudices; to be blunt, that much of social science isn't worth the candle and is disappearing up its backside.

That uncomfortable analysis sits behind what I have attempted to do in this book. I remain deeply committed to critical social science and to its importance in the wider academy and to societies. But what I find myself increasingly exercised by is the need to demonstrate just why and how the social sciences matter. How might that be done? My answer to that question is: 1) to produce analyses that focus on what others see as identifiable problems – not to respond to the current milieu by retreating, to concentrate on 'island' activities, or the topics and debates that we think are interesting; and 2) to use that analysis to create clearly understandable messages that can be put to work in other arenas, particularly by policymakers. We need, in short, to write social science not for ourselves but in ways that reach out and allow others in.

What I have written here, therefore, is a book with those maxims at its heart. It is a book that has been designed with my learning from the policy-writing experience front and centre. It therefore works as both a framing exercise for understanding the global waste problem (Chapter 1) and as key messages and their implications (Chapter 7). In the middle (Chapters 2–6) is the social science research that informs that frame (the evidence, in these terms) and that underpins the messages. While it is a book that certainly can be read by academics, researchers and students, it is a book that is not aimed at just them. I hope it is much more widely accessible. That is reflected in the style I've adopted, which is purposely less academic. So, as well as being written in more everyday language, there are no tables or graphs, neither are there references in any of the text, and the welter of research on which the book is based is only discussed in the notes.

What follows in substance, then, is a synthesis with the global waste problem at its centre, but I use critical social science perspectives to understand that problem. Doing that has required me to revisit my earlier work on consumption. This has not only necessitated some updating, in the light of the rise of digital media and platforms and consumption on the go, but also required extending it, such that it incorporates a stronger connection with economic work. I then put that revised and updated work into dialogue with my more recent research on waste, synthesizing across multiple papers written in the course of the last 15 years but also being unafraid to travel in directions new, where the focus requires it. In such a way, the analysis ventures far from the core terrain of critical waste studies[13] to encompass not just the technical waste management literature, but also the technical literature on waste-based innovation and work on financialization and the decarbonization agenda. The result is a challenge to current policy framings of the global waste problem. In that regard, my hope is that the book shows what critical social science perspectives do best. But it is also a book that, in its conclusions, offers policymakers routes forwards, ones that show that if we are to make inroads into the global waste problem, then we need to be looking in other directions than waste.

Nicky Gregson
August 2022

1

The Global Waste Problem and How to Think about It

Or, how to understand the 'too much waste' problem

That the world has a waste problem is incontrovertible. Equally obvious is that this problem connects to global levels of consumption. What is debatable, however, is how those connections are drawn and understood. The arguments that are made and then developed in this book are born of the contention that the ways these connections have been drawn are fundamentally flawed. More than that, because they are flawed, the interventions that they suggest are awry. While they may be motivated by the very best of intentions, many of the interventions designed to address the problem of too much waste actually have the opposite effect: they work as incentives to generate more of the stuff. Even worse, many of the solutions put in place to manage waste demand that more waste be produced. This, I would argue, is down to how we understand and think about waste, its relation to consumption, and the relation of both to economic activity.

Before I begin to open this up, let's take a look at the scale of the waste problem. Some of the numbers here are mind-boggling. They evidence that the world most definitely has a waste problem.[1] Take construction and demolition wastes, for example. These wastes are closely related to rates of economic growth. Booming economies are characterized by a huge expansion in construction activity, not just in relation to residential uses but also industrial premises and corporate facilities. Construction and demolition wastes are sizeable contributors to the total wastes in every country.[2] An estimated 3 billion-plus tonnes of construction and demolition wastes are generated annually globally, much of it concrete, which ends up mostly landfilled. China alone accounts for over 1 billion tonnes of these wastes.[3] Undoubtedly, construction and demolition wastes are a huge part of the global waste problem, but, like many of the wastes that result from industry, they are niche; they only trouble those who work in the construction and demolition industries and regulators.

It is a similar story with a very different type of waste: radioactive wastes. These wastes are an unavoidable effect of nuclear power generation and are an inevitable part of a future powered by non-fossil fuel-derived energy. So, a solution to the radioactive waste problem is imperative to find. Although it is the most radioactive of these wastes that attract the most attention, and that pose the most obvious political and technological challenges for radioactive waste management[4], the largest volumes of radioactive waste comprise quite ordinary materials that are contaminated through being put to work in the nuclear industry, for example, protective clothing and the rubble from building materials.[5] These materials constitute some of the biggest challenges for the nuclear decommissioning industry, for their projected future volume is an order of magnitude greater than higher-activity radioactive wastes.[6] Indeed, projections show that low-level radioactive wastes threaten to overwhelm the capacities of the industry's designated low-level waste repositories.[7]

Construction and demolition waste and radioactive waste are out of sight and out of mind for most people. This makes them of little wider concern. It is the same with all kinds of industrial wastes, which are the major contributors by volume to the global waste problem.[8]

But, the same issue of massive numbers can also be found in relation to consumer-derived (or municipal) waste – the type of stuff that is everywhere in most people's lives and that they certainly do care about. One of the best examples of public concern about such waste in recent years is plastic bags. Again, the numbers involved are eye-watering. In 2007, global estimates put the number of bags being consumed and discarded per year in the range of 500 billion–1.5 trillion, with the biggest generators being China (300 billion–1 trillion) and the US (100 billion).[9] Other examples of particular types of municipal waste that have captured public attention in recent years in the UK are disposable nappies and food packaging. The lightness of plastic means the contribution of plastic bags and food packaging to the total weight of global waste is tiny. But these are the kinds of things that elicit concern from the general public. In democratic societies where the primary goal of the political class is re-election, voters' concerns trump everything else. So it is that while construction and demolition waste and radioactive wastes are vastly more important components by weight of the UK's waste problem, it is interventions in relation to plastic bags and packaging that have come to pass.[10]

Paying attention to the amounts of various types of waste being generated is obviously important. It is not my intention to try to argue otherwise. But, while this works to establish and evidence a problem, it's not necessarily a helpful way of thinking about the global waste problem. Indeed, I would argue that there is a need to be mindful of thinking about waste in weighty, categorical terms. This is because although the source of waste is recognized in these numbers (as the construction and demolition industries, the nuclear power generation industry, consumers), what ends up mattering is the amounts of different types of waste and the relativities of those amounts by different categories. This has effects. Chief of these is that the role of everything else in understanding the generation of these wastes is bracketed out, or left tacit and mostly unexamined. The potential, then, for what scientists and engineers often label as the 'treat the symptom not the cause' situation to emerge is high.

As I now show, through a focus on first the policy domain and then the activist domain, bracketing out makes tacit a whole host of assumptions about waste. These assumptions need to be made explicit. As I will show, they matter profoundly to how waste is being understood in these contexts, to the formulation of the global waste problem and then to what type of interventions get suggested.

★ ★ ★

To begin to open this up, let's take how wastes are thought about in policy circles. There are multiple levels here, of course: global, international, national or regional, or one that accords with the finer levels of granularity that comprise local government: cities, towns and smaller settlements. But the general approach across different scales and jurisdictions is by and large consistent. Development and growth are seen to connect positively with an increase in consumption (more consumption − increased growth), while rising levels of consumption are seen to lead to rising amounts of waste. Waste is typically measured by weight and volume and characterized by type − a categorization exercise that reflects both material composition and/or source. So, we find that the global waste problem is represented in the policy domain as one of rising amounts of e-waste, construction waste, plastic waste, food waste, consumer waste, packaging waste, radioactive waste, mine waste, and so on. One of the best and most influential examples of this approach comes from the World Bank. It serves to demonstrate the scale of the global municipal waste challenge and why this is exercising the minds of policymakers.

A highly influential World Bank report from 2012 estimated a global population of 3 billion living in cities then generating 1.3 billion tonnes of waste. It also gave projections through to 2025, by which point it was estimated that there would be 4.3 billion people living in cities producing 2.2 billion tonnes of waste.[11] Updated in 2019, the latest World Bank report, based on data from 217 countries and 367 cities, confirms these projections. It suggests 2 billion tonnes of waste were generated in 2016 and projects this growing to 2.59 billion tonnes by 2030 and 3.4 billion tonnes by 2050.[12] Disaggregated by global regions, that 2012 data for the East Asia/Pacific region recorded 260 million tonnes of waste being generated annually (70 per cent of this in China), with the 2025 projections for the same region increasing to 1.86 billion tonnes, greater than the projection for the Organisation for Economic Co-operation and Development group of countries (1.7 billion tonnes). The 2019 report showed that by 2016, that 260 million tonnes had already increased to 468 million tonnes and accounted for 23 per cent of global municipal waste. Projections for 2050 estimate a 19 per cent increase in annual municipal waste across the high-income group of countries, but they also suggest that waste generation in lower- and middle-income countries will grow much more rapidly. This is forecast to grow by 40 per cent, with the most rapid increases anticipated in Sub-Saharan Africa, South Asia, the Middle East and North Africa. Although the 2019 data does not disaggregate by cities, the earlier 2012 report gives an indication of the size of the challenge facing individual cities. At that time, the megacities with populations estimated at greater than 10 million, such as Beijing, Shanghai, Buenos Aires, São Paulo, Delhi and Mumbai, were generating between 2.0 and 7.6 million tonnes/year of municipal waste, with the highest figures being for the South American megacities. A similar picture is to be found in the major cities of the Middle East and North Africa, where 2012 estimates for Cairo, Tehran and Baghdad were 5 million tonnes, 2.6 million tonnes and 4.2 million tonnes, respectively.

Framing global municipal waste in this way leads to understanding it as weighty stuff. It also leads to an understanding of the global municipal waste problem as one of 'too much waste'. At the same time, there is an imperative to act that comes with figures like these, for the strong suggestion is that, unless they act, certain cities will find themselves overwhelmed by a tsunami of waste as an effect of rapid economic growth. Framed in this way, the imperative to governments and policymakers is seemingly obvious: reduce waste, and fast. That imperative, as we will see in Chapter 6, is significant. It has affected what type of waste management solutions are being put in place across large parts of the developing world. But it is just as important to note what is tacit here – that the role of growth in all this goes unchallenged. There is nothing in formulations such as these that leads to any questioning of the necessity for continued economic growth. If anything, it's the reverse: solve the waste problem and economic growth can continue untrammelled, with no effects. To find that kind of argument here is not surprising. These are World Bank reports, after all, and the World Bank is one of the global institutions whose task it is to perpetuate and encourage the conditions for growth. But those assumptions certainly do need to be made explicit. Let's explore this a bit further by considering how the connections between consumption, waste and economic activity are being drawn here.

In so far as the relationship between waste and consumption is concerned, the chief effect of leaving growth unchallenged is that the 'too much waste' problem isn't so much a problem *of* consumption; rather, it's the waste that results *from* consumption that is deemed the problem. This allows a sleight of hand – consumption's role in waste generation can now be side-lined and placed in the metaphorical parking lot. This is mightily convenient. It allows governments and policymakers to continue to take as read that the role of consumers in any economy is to keep on consuming, that is, buying goods and services. Simultaneously, it also props up one of the core assumptions of global economic policy, that it is generally a good thing for more of the world's economies to become increasingly reliant upon consumption as an economic activity. So, this way of thinking encourages more countries to pull various policy levers that in turn create and encourage the conditions by which more of their populations have incomes sufficient to enable the purchase of consumer goods; and then, when they have reached that level of economic development, intensify that by getting more people to buy more goods more frequently. Decoupling waste from consumption, then, allows for perpetuating the idea that we can all be global consumers. With consumption side-lined yet sacrosanct, the task then becomes to address the connection between waste and growth. The policy goal becomes growth without waste or, at the very least, growth with reduced waste. There is an assumption, then, that growth and waste can be decoupled.

So how is that to be achieved? Having formulated the solution to the waste problem as one of waste reduction, how have policymakers operationalized that? To answer that question we need to go back to the weightiness of waste. When waste is understood as (too much) weighty stuff (of various categories), then the obvious answer to the question of how to reduce waste is to decrease the weight of waste that is generated in these categories. Cue a world of complexity, for while weight is an absolute scientific measure, what counts as waste most certainly isn't.

What counts as waste is firmly located in the terrain of what is known in the social sciences as the politics of measurement. The most notable, and probably the best, recent example of this is the number of deaths recorded within individual countries as an effect of COVID-19, where governments of all stripes have engaged in multiple games as to what counts as a COVID-19 death. Why have they done so? Because charts comprised of death rates are open to comparison; measurement therefore becomes a means to benchmark the success (or otherwise) of various interventions. Measurement then feeds through into performance indicators and ultimately into a political scorecard – and failures accrue and have consequences. In democratic societies, failures become the means for opposing parties to challenge the record of incumbent administrations, and so are seen to be potential vote winners and losers. So, all sorts of games get played over measurement. And games most certainly have been played with respect to measuring waste reduction.

If I were to characterize this game in Europe, where reducing waste generation has been a strong policy goal over some 20-plus years now, it would be something like 'chase the tonnage'. It goes something like this. Way back in the 1990s, when efforts to reduce the amount of waste being generated in the EU were in their infancy, the amount of waste then being generated by each member state was set as a baseline, against which all reduction efforts were to be assessed. Five-year targets were then set against those baselines, with each target being a ramp up on the previous, while penalties – in the form of fines – acted as the incentive for member states to get their waste house in order. But what exactly counted as waste in that original baselining exercise? The answer was the stuff that ended up in landfill. In such a way, we see how what counts as waste is always a political matter. For while there are a number of very good reasons why landfill is not a good option for managing a lot of wastes – most notably its contribution to greenhouse gases (GHGs) via methane generation and to groundwater pollution via leachates – it was not clear, particularly to member states that relied on landfilling as a technology, why it was that much worse a solution to waste management than incineration. Waste management solutions, then, and the favouring of one technical option over another, mapped into political differences between EU member states, with landfill-dependent states losing out to those dependent on incineration, the upshot being that they were required to undertake a radical (and costly) overhaul of how they managed their wastes. At the same time, that radical overhaul had to demonstrate waste reduction but, and this is the critical bit, that reduction was operationalized in measurement as diversion from landfill. Waste reduction then became framed not as waste reduction per se but as material saved from being landfilled. It was at this juncture that the game of 'chase the tonnage' really began, for weight is a much easier form of measurement to demonstrate diversion from landfill than quantifying GHG savings or groundwater pollution. It is also the baseline unit of measurement in the waste management industry. All manner of collection schemes targeting the weightier types of material in the municipal waste stream were therefore introduced, for example, glass and paper. No matter that much of this material actually ended up being exported to China and countries in the global South (where it has long been seen as 'foreign garbage'), the point of the game was to capture material to be weighed and therefore to count as material diverted from the EU's landfills. And the game has continued. For example, in the UK, one of the categories of materials that counts towards recycling targets is 'garden waste', in other words, grass cuttings, all manner of plant matter and hedge trimmings. Organic matter like this – the type of stuff that is used to produce compost – is not what many would probably first think of when they think of post-consumer, municipal waste. But because they are no longer being landfilled, they count towards waste reduction, and because this stuff tends to weigh a fair amount, it's valuable to collect and count in to that total. Hence why many local authorities in the UK see a point to running garden waste collection services, rather than issuing composting bins to households with gardens.

The 'chase the tonnage' game begins to introduce the world of perverse incentives that characterizes so much of what purport to be policy interventions designed to reduce waste. It shows how what this encourages is collecting, or 'harvesting', more material. As consumers, businesses and organizations have been issued with more and more bins for more and more categories of materials, and as the content of those bins are collected, amalgamated and then weighed, so we can see how we are all captured in a game that encourages people to discard more and more material while simultaneously presenting that as reducing waste. It is genuinely Alice in Wonderland stuff; or, a hall of mirrors. Yet, what it does, or what it ensures, is that people carry on consuming, for if we didn't, these bins wouldn't be full, the habit of putting the bins out wouldn't make any sense and their contents wouldn't be worth the expense of their collection.

★ ★ ★

At this point, I want to turn to a very different strand of thought: the kinds of arguments that are made by waste activists and campaign groups.

As in mainstream policy circles, for activists the global waste problem is largely a problem evidenced by the weightiness and sheer amount of waste. But here that goes beyond the facts of this matter; it is taken as a manifestation of the wastefulness of modern life and modern lives. It is also a problem that activists connect to social injustices. So here, waste doesn't just remain waste, it also becomes pollution – the stuff that is dumped into open environments such as watercourses, drains, by roads, into the sea, and that then circulates, affecting adversely all the lives (human, animal, plant, bird, fish) it comes into contact with. So in activist circles, waste isn't just a problem, it's actively seen as 'bad'. That means that it is not just seen as a problem needing to be reduced; rather, it is seen as something that can and should (the moral imperative is always important here) be prevented. In other words, waste is a category that should be avoided as much as possible. As we will see, this has considerable implications in terms of the interventions that are suggested.

Before we consider these, it is important to clarify the differences between reduction and prevention. Reduction, most definitely, is not the same thing as prevention, and the two should not be confused. Whereas waste reduction policies try to reduce the amount of stuff becoming waste, interventions aimed at waste prevention seek to stop stuff from becoming waste at all. Rather than less waste making, this is about saving stuff from becoming waste in the first place. It tries to eliminate the waste category.

Activist-led interventions that seek to save stuff from becoming waste typically revolve around two principles: extending the durability of human-made stuff and eking out the use to which stuff that has already been made is put. The two are interlinked, for the underlying premise of both is to maximize the utility of that which has already been produced rather than satisfying needs and desires by creating more stuff. So we can see that here there is a fundamental challenge to the 'business as usual' economic growth model that prevails in policy circles. Instead, this is consumption minus growth, or consumption premised on what is increasingly being represented as an 'economics of enough'.[13] So far so good – after all, there is much that is attractive about an argument that recognizes a fundamental cause of the global waste problem is the global economic model itself. The difficulty, however, lies in designing interventions that challenge that. This is where things start to unravel. As we will see, much of this comes down to another version of bracketing out, in which moral arguments about how things ought to be gloss over the disjuncture between that and what actually goes on by way of consumption in homes, and why. In this case, then, what gets bracketed out is consumption. The effect is that what gets overlooked is what it would take by way of societal change for particular seen to be desirable waste-saving changes to occur.

Let's open this up by focusing to begin with on the interventions that are seen to be waste saving. Alongside a long-standing emphasis in green circles on repair and reuse rather than substitution and replacement, one of the concepts gaining increasing traction in waste activist circles currently is the notion of the surplus.[14] There is much that is attractive about this idea, and I have certainly argued previously that keeping things in the category of the surplus is one way to stop stuff from becoming waste.[15] I still hold to that. Moreover, as I will show in Chapter 4, the concept of the surplus is extremely helpful in aiding thinking about consumer discard. The core point about the surplus as this pertains to waste saving, however, is one that harks back to its origins in agricultural societies: it is a means of using redistribution to deal with the glut that can often come with harvests. So it is not an everyday category, at least with regards to waste saving. But some of the arguments and interventions that are now being made to save stuff from becoming waste seek to operationalize the surplus at scale and in an everyday capacity, typically by working with digital platforms (chiefly apps) to connect those with surplus stuff with others who might want or find a use for this stuff. I will look in more depth at this in Chapter 4, using Freecycle as our example, but another good illustration of the same principle occurs in relation to food waste.[16] In what follows, I use food waste as an extended example, both to show how this is being connected to the idea of the surplus and to begin to outline how waste-saving interventions bracket out consumption. This is then followed by a preliminary exploration of how opening up consumption allows us to see food waste differently.

Unpacking food waste

There is no denying that food waste is a massive issue, a significant part of the global waste problem and one laced through with moral conundrums. The disparity both in the amount and levels of food poverty between rich and poor nations has long been an important component of this debate, but growing inequalities in access to food within rich and middle-income countries bring this debate closer to home and sit uneasily with data that shows vast quantities of food that is produced for human consumption being thrown away by members of the same society.[17] There is, after all, something that borders on the obscene in the paradox that in a country like the UK where millions are dependent upon food charities, millions more continue to throw edible food out.[18] But the UK is far from alone in this in the countries of the global North. Within the EU, an estimated 88 million tonnes of food is wasted annually, the majority of which comes from households and the hospitality sector.[19] In the US, a 2010 estimate put this figure at around 66 million tonnes.[20] In China too, rising amounts of food waste, particularly emanating from the hospitality sector, means that food waste is rising up the political agenda.[21]

So how does the idea of the surplus relate to food waste? To unpack this, we need to think about what the surplus means. Let's start in the context of the food retail and hospitality sectors. Like any other business situation, 'surplus' food here represents an excess of supply over demand. But it is also an effect of business economics, in which the impossibility of matching supply and demand precisely results in the practice of overordering. For retailers in general, overordering is vastly preferable to underordering, for selling out of a particular product line means lost sales and lost revenue. By contrast, provided stock sells, overordering can be easily accommodated, as the costs of small amounts of unsold stock can be factored into pricing to the consumer. In grocery food retail, however, such is the volume of stock and the degree of market concentration that the effect of overordering in physical goods is that significant amounts of unsold stock are left over in the major supermarkets on a daily basis, most especially, fresh, chilled and/or ambient, and short shelf-life products. This is the supermarket surplus and, outside of the 'deep discounters' such as Aldi and Lidl, it is a direct result of offering an abundance of consumer choice. It is a similar situation – at a reduced scale – in the hospitality sector. Here the surplus results from: 1) the mismatch between the number of 'covers' (or meals) that are actually ordered and the number of meals that the restaurant estimates as needing to be produced (booked tables); and 2) surplus stock, for example, when restaurants buy slightly more than they need of certain ingredients to ensure that there are sufficient stocks to cope with the eventuality where things go wrong in the kitchen.

Managing the surplus in hospitality is difficult. The food is often fresh and/or chilled and so has a short use window. It can often be high end and more niche, and so difficult to repurpose to make into the more standard dishes that are typically offered in the charity food sector. Given the hours of business, this surplus often emerges late in the evening, making it difficult for many third-sector organizations to collect and process. This is why hospitality is a major source of food waste. By comparison, supermarkets have slightly more room to manoeuvre. One response of supermarkets to the surplus is the end-of-day heavy discounting of goods coming up to their best-by or use-by date. Supermarket staff will often take advantage of this at the end of shifts; so too do savvy consumers who shop towards the end of the trading day, but they cannot be relied on to shift all surplus stock. So, what remains is channelled towards the charity food sector, mostly via partnership agreements with specific organizations.[22] Surplus food – which, by definition, is unpredictable[23] – gets connected up with third-sector organizations, which function either as warehouses and distribution centres ('food banks') or as meal providers for benefit-dependent and, often, homeless people. Managing the surplus in this way then is a means to societal redistribution; it addresses poverty and inequality. But what we also should note is that in treating the symptom (unsold food), not the cause (business economics), it allows supermarkets to continue with 'business as usual' while benefitting from the moral high ground that comes with food redistribution. In such a way, using the surplus to save food from becoming waste also becomes a means to saving the supermarket model of food provisioning.

What about ordinary consumers? How does surplus food emerge from households? One of the major transformations in food provisioning over the past 20 years has been in the reduction in the amount of food prepared in households. We see this in the expansion in 'eating out' and the more recent rise in on-demand meal delivery services. Mostly, though, ordinary households still make do most of the time with self-provisioning, or unwaged labour, to feed themselves. Since this type of food provisioning is outside the market, it is entirely beyond the supply–demand relation. So, if we are to understand how the surplus is derived here, we have to look at what is going on within consumption.

At one level, there is no getting away from it that a surplus that is derived from the household has to result from over-acquisition and over-provisioning. It can result from buying more food than can be consumed by the household in any given period, taking use-by dates into account, and/or from putting too much on meal plates. This is the phenomenon known as 'overconsumption', where the stock of goods held exceeds the capacity to use them up and it results inevitably in a system that overflows. Getting rid of surplus food via an app, or throwing it out or binning food that is left uneaten after a meal, is a way of externalizing that overconsumption – it's a way of dealing with waste.

Identifying overconsumption in the household is one thing; quite another is accounting for it. We will explore this issue and some of the reasons for it in depth in Chapters 2 and 3, but fundamentally, if we are to really understand this, we have to delve much deeper than is typical in waste activist circles into what is going on in within-household consumption. For sure, some of the explanations for household food waste that are offered by activists and campaigners undoubtedly do come into this. This does connect to how food is sold by supermarkets (for example, cheaper, larger packs, multibuy offers). Some of it is down to patterns of shopping (weekly, in bulk), and some of it is about what are labelled as 'cultures of consumption', which, in this instance, get identified with the loss of thrifty cooking practices, or the arts of cooking with leftovers and remainders, and of eking out certain foodstuffs such that they will make multiple meals over days rather than just one meal. But note here the tendency to focus on solving the symptom of the problem. This is all about trying to stop food becoming food waste, not about trying to understand why food waste gets generated in the first place. So, we find campaigns to get supermarkets to sell food in smaller quantities (fewer 'family' packs, more individual portions). Then there are all those handy hints as to how to shop in ways that will produce less food waste: make lists, use an app when shopping, plan meals, cook ahead and use the freezer. And then there are a host of 'how to' interventions that seek to turn us all into being dab hands at cooking with leftovers or with the type of stuff that often ends up as food waste, for instance, stale bread. But while these interventions have made some inroads into household food waste generation, they have not been nearly enough. Why? There is one primary reason that I would identify above all else. This is that, in common with many other activist groups, food waste activists and their campaigns bracket out what social scientists refer to as 'the social', understood as the underpinning order and ordering of society that shapes the day-to-day patterns and rhythms of how we live. They are, in other words, sociologically weak. So let's begin to explore what a sociologically richer understanding of food waste looks like.[24]

To drill down into this and most especially its effects, let's take the ordering part of this first. One of the things activist perspectives overlook, largely because of their focus on present and future change, is the importance of fundamental transformations in the organization of societies. Typically, these are the sorts of changes that are enacted over decades – think of them as social 'slow burns'. Because activist perspectives don't tend to acknowledge this sociological bedrock, they don't tend to ask how socially and economically possible the sorts of changes they are advocating are. Rather, they tend to assume that everyone will see the world through their eyes. This is the activist blind spot. Neither, when such perspectives look back to old ways of doing things, such as thrifty forms of meal preparation and cooking, and urge their return, do they ask what kind of societal organization made it possible to do these. So, in turn, what gets overlooked is a huge political question, which is: what would it require of people living today to adopt such changes? And then, is that kind of intervention even realistic or doable?

If we home in on the UK context, we'll get some idea of just how much is at stake here. Behind the posing of these questions lie two of the profoundest sociological changes of the 20th century – changes that have radically restructured the ordering of British society and that most people would probably regard as irreversible. These are: 1) the growth in women's employment and the allied wider feminization of economic life; and 2) the changing relationship between humans, food and different categories of animals. Let's take each of these in turn.

Up until the 1960s in Britain, with the exception of World War II, there was a clear gender division of labour: while men's role was to be the household 'breadwinner' (a role that necessitated being in paid employment and that had obvious difficulties if one was not), women's place was in the home. Women's work therefore primarily entailed socially reproductive labour, not only childcare but all forms of domestic work, including shopping, washing and ironing, cleaning, meal preparation, cooking and baking. Women's entry into the labour market from the 1970s not only transformed the world of paid work, it also transformed social reproductive work, and the relationship between these two spheres of work. With women increasingly out of the home during working hours, domestic work became mediated more by technology and the market. But it was also ordered increasingly by convenience.[25] Fundamentally, convenience is about time. The sociological literature highlights two ways of thinking about this. The first is the standard sense, of a means to addressing time shortages by saving time – think of appeals to 'labour-saving' devices or the development of appliances with delayed timers. The second is to recognize that time is also about space; or, when we're talking about convenience what we often mean is introducing interventions that address the complexities of time-space coordination, most of which stem from the impossibility of physically being in more than one place at once. Sociological research points to the complexities of time-space coordination, sequencing and shifting in contemporary social life. It highlights the speed, intensity and endless 'busyness' of most people's everyday lives. As I now show, if we are to understand household-generated food waste, we need to understand how convenience in both senses of the word makes food waste.

Both senses of convenience are everywhere in relation to food. The rise of eating out, of the 'takeaway' and more recently of app-mediated meal ordering and delivery services all show how the market has succeeded in cashing in on appealing to saving time on food shopping, meal preparation and cooking. Inside the home, too, the increased ownership of certain devices, notably fridges, freezers and most especially microwaves, is indicative not just of the appeal of time-saving but its necessity. Indeed, most households would probably see all three as essential devices for managing the tasks of food storage and preparation. Then there is the supermarket, which, as well as offering the consumer choice, specializes in selling convenience to consumers. The growth of supermarkets in the second half of the 20th century, and their subsequent dominance of the grocery sector, is not just down to economics. It is in part explicable by their success in selling 'time-poor' households[26] both convenience shopping (the weekly household shopping trip/delivery rather than shopping daily) and convenience food and meals. Convenience shopping is hard-wired into supermarket shopping – look at the trolleys that, in their size and design, script the purchase of large amounts of food and therefore push people towards shopping as a once-a-week, car-dependent, bulk-buying act.[27] Then there is the food on sale. Many of these products come ready-made. They are sold to consumers as time-saving food that can be stored in the fridge or freezer and that just needs to be zapped for a few minutes in a microwave before being eaten; or, it's food in a tin, which just needs a blast of energy (gas or electricity) to turn it into food ready to consume.

As well as having effects on technological innovation and technology adoption within the home, I would suggest that selling convenience has also had major effects on the contents of household bins and that it is this convenience rather than wasteful consumers that we need to see as generative of food waste. The turn to convenience has seen a decline in food preparation and cooking from scratch and a decline in thrifty cooking practices. Thrifty cooking, even if one has the time to do it, is just that much harder to do with convenience food. This is because it starts from the leftovers of a ready-made meal, such as a half-eaten lasagne, rather than stretching out the basic ingredients over days.[28] Then there is the effect of convenience shopping. This manifests itself in the combined effects of bulk purchasing and use-by dates. Waste audit after waste audit has revealed this: stale or mouldy bread, over-ripe bananas, yoghurts and salad bags past their use-by date, sprouting potatoes, and mouldy fruit and vegetables figure strongly in what is thrown out as food waste in the UK.[29] Such deemed to be inedible food is in part an effect of the ways in which supermarkets pass on the inevitable decay and deterioration of food to the consumer, but it is also indicative of how time-space coordination and sequencing relate to food provisioning. At the most basic of levels, food provisioning (encompassing shopping and preparation/cooking) has to fit in with everything else that is going on in a household. In households that include children and teenagers, that is a lot of activities that need scheduling and coordinating: football training, matches, dance classes and competitions, drama, music lessons, sleepovers, multiple birthday parties, and that is without the visits to grandparents, trips out and holidays. Convenience shopping – as offered by supermarkets, be that through the in-person trip or the one-hour home delivery service – enables that kind of fine-grained coordination and scheduling. It's a mode of shopping for a modern social life. But convenience doesn't track through in quite the same way to food preparation, which remains a juggling act for most households – something that has to be fitted in with all the other competing demands. The effect is that, notwithstanding the best of intentions, food purchased can all too easily fall out of the category of what is deemed to be safely edible. The effects of that, in turn, are visible in the contents of the bin.

Also visible in the food waste content of UK household bins is a rather different profound transformation in the categories that order human life. This is in how we live with companion animals (or, our pets) and most especially in how we feed them. Here, too, the effects of convenience in food waste are there to be seen.

The waste activist literature often makes reference – much of it approvingly – to the potential role of domesticated farm animals (livestock) as bioprocessors of inedible or unwanted surplus human food. Turning farm animals into bioprocessors is seen as a means of turning discarded human food into meat and dairy products. It incorporates livestock into the ideal of the circular economy. Given their history, much of the interest focuses on pigs, which were domesticated around the 9th millennium BC by being fed leftover food and scraps. Even as late as the 1960s, those practices were still common in the UK in rural areas and on the urban periphery, but they then went into steep decline. The demand for cheap meat required intensified methods of livestock production. Together with the implementation of food standards and safety, this led to livestock feed becoming commercialized and primarily manufactured. It led to the development of a huge agri-food sector. So the activist argument goes, and much as with the appeal to rediscovering old thrifty ways of cooking, one way of attending to the global food waste challenge is to rekindle those old practices of pig husbandry. The counter argument raises points of biosecurity.[30] Less recognized in all this debate is the lost role of companion animals as consumers of discarded or surplus human food.

At the same time as the rise of women's employment brought about a transformation in the gender division of labour, a parallel transformation went on in how humans live with companion animals. This is manifested in how we feed those animals. To sketch this out it is useful to backtrack from the present. In the latter part of the 20th century, companion animals (dogs and cats chiefly) started to be seen as pseudo-humans; they have now become fully-fledged household members rather than appendages to a household. Nowhere is this personhood status clearer than at a veterinary surgery, where the animal is known (and addressed) by its pet name and then the owner/client's surname. This is a clear departure from my first memories of trips to the vets, where cats were simply cats (a species of animal) and not even referred to by a pet name, let alone given personhood status.[31] But even then, back in the 1960s, those cats, like the humans, were being fed, at least in part, through convenience. Just as we were fed via tins of prepared ready-to-eat food, so too were they – in their case, with tins of Whiskas® rather than baked beans, spaghetti and spam! Even though those cats supplemented that diet with a regular bird or two, their diets had been aligned, through convenience, with human diets. It was the same for dog-owning households, although here Whiskas was substituted by Chum®. Go back just a decade or so though and in the 1950s, and certainly the decades preceding, feeding the family pet most definitely did not look like this. Whereas cats may have been given access to the occasional scraps, most were expected to fend for themselves, catching birds and mice. Indeed, that was why they were tolerated – as a means of rodent control. Dogs, by contrast, and in a pattern that goes back over millennia to the original domestication of scavenger wolves, were fed with the surplus from human food consumption. The 'dog's dinner' (a phrase that remains in everyday use[32]) was literally a plate of piled-up scraps left over from the meals eaten by the human members of the household that day. Fast forward to today and these same companion animals don't just have single supermarket aisles dedicated to their food needs; they have entire supermarkets, supplying a vast array of bags, pouches, sachets and tins of manufactured foodstuffs (dried and wet), toys and care-related products. Look closely at the content of their thoroughly commodified care and it clearly demonstrates their status as person equivalents. But just as convenience-fed humans generate food waste, so too do our convenience-fed companion animals – in half-eaten bowls of tinned food, tins that won't be eaten and dried ('special diet') food (for the older pet and to manage pet obesity) that is left untouched. From being the original scavengers who lived off the food discarded by humans, convenience has turned our companion animals into waste makers in our own image.

★ ★ ★

Having begun this chapter by outlining how the global waste problem is typically framed and the problems that result from that framing, I now move to the approaches and perspectives that inform and shape this book. My starting point is simply stated: that it is impossible to understand the global waste problem by beginning with waste. If we do that we start with the stuff itself and that, as we saw with the example of food waste, can be highly misleading. So much so that we end up suggesting interventions that treat symptoms without addressing causes. A better tactic, I propose, is to begin by acknowledging that the 'what' of waste (or what we see and label as waste) is always an effect. Or, to use a phrase of Erich Zimmerman's, that 'wastes are not; they become'.[33] If we are to understand waste, I maintain, we have to understand its generative processes.

That said, a degree of ring-fencing is necessary. Wastes cannot only be anything but they are also everywhere in human life, as will become clear in Chapter 2. That is not only unhelpful; it's unmanageable if we are to retain any degree of focus. So, which wastes should we focus on, and why? Notwithstanding that industrial wastes constitute the major part by weight and volume of global waste, my primary focus in this book is on consumer-generated wastes, that is, municipal waste – or, the kind of stuff that you and I discard from our homes.

Consumer-generated waste is what defines modern waste as different; it is what sets us apart from the wastes of our ancestors. Go way back in time and academic research has demonstrated that human societies have always generated wastes from what we might loosely call productive activities as well as from basic consumption – or, consuming the stuff that keeps us alive. This we can see from archaeologists' excavations of the middens and deposition pits of prehistoric and early medieval societies. The midden at Taperhina is a good example. This is an early pottery-age fishing village in the Brazilian Amazon. Its midden is a mound over 6 m high, covering several hectares. Like all prehistoric middens, it contains pottery shards along with broken flakes from stone-age tools and the inedible remains of food, such as mussel shells and fish bones.[34] That balance between different types of deposition – that resulting from food consumption and the residues of productive activity – is how things pretty much stayed over further centuries of human history, until the birth of industrial capitalism. Even then, while industry expanded and industrial wastes (particularly mining wastes and the wastes from manufacturing such as from steel and chemical production) presented major problems of environmental pollution, the wastes generated from everyday living remained limited to the residues of heating (ash chiefly), food preparation and eating. As we will see in Chapter 3, all that changed in the aftermath of the Great Depression in the US and Europe. From that point onwards, consumption – as in the purchase of goods and then services – became of critical importance to economic growth. In turn, sustaining growth came to depend on the acquisition of stuff, and not only this, endless cycles of the replacement of stuff. Populations that until the 1930s had been seen primarily as supplying a workforce to support the requirements of industry now became seen as just as important for their role as the consumers of goods and services. The effect was that those same populations started to generate more and more of what we now term consumer, or municipal, waste. The relationship between consumption and waste came to be mediated economically, and it is a relationship that works through the conditions, or context, of particular economies.

All this has implications for how we might understand consumer-generated waste. I will state this first as three central premises and then elaborate on each, before outlining how they map into the chapters of the book. The premises:

1. To understand consumer-generated waste, we have to understand consumption – not see waste as an automatic, inevitable effect of consumption.
2. To understand consumption requires us to understand what's going on when people get rid of things. It's not sufficient to understand consumption as the acquisition of goods and services. We also have to understand how this connects to the use of things, to the devaluation of things and to their disposal.
3. All of this is economically mediated. This means not just that it is shaped by the differential realities of living in particular conditions in particular economies (the different people living in different circumstances in different places will generate different types of consumer waste issues), but also that consumer waste itself is subject to economic forces. Consumer waste is no longer an economic externality. It has been economized; it is thoroughly bound up in the shape and dynamics of contemporary capitalism.

Premise 1

For too long, understanding of the relationship between consumption and waste has been proceeding without recognition of the advances that have occurred in consumption research in recent decades. Perusal of the pages of the mainstream media, or an engagement with the policy and activist fields, will very quickly show that contemporary levels of waste (and therefore the waste problem) are connected to consumption as *consumerism*. One is seen as evidence of the other; or, consumption is understood in terms of waste. In such a way, acquisition gets read as the desire for goods and as the endless quest for the new, which in turn results in never-ending cycles of discarding and waste. Consumerism is connected to chucking out stuff; it is seen as a manifestation of an endemic wastefulness, often encapsulated in terms such as 'the throwaway society'. This may have been the understanding of consumption that prevailed in the academic literature of the mid-20th century but it is now not just outmoded (if it ever was actually accurate), but well off the mark. The best part of 20 years of research has advanced understandings of consumption no end. These understandings need to inform how connections are made between consumption and waste.

It is, of course, impossible to do justice to the full range of this now enormous literature here, but some strands of study have been particularly important to debunking the myth of consumerism. Running across these has been a shift in the locus of research: consumption research shifted from what had hitherto been its primary site in retail spaces (principally then shops and malls) to crossing the threshold of the household to enable the study of how acquisition related to actual consumption in and out of people's homes.[35] That shift proved critical in opening up a new wave of studies of shopping. Rather than assume what was going on when people went shopping, researchers actually went shopping with people, and what they found was little evidence of a rapacious desire for the new but rather a much more mundane world of ordinary provisioning that showed the connections between shopping and key social identities.[36] Consumerism was nowhere in sight. So, when people went shopping (particularly food shopping), they turned out to be not just buying for a household but enacting their primary social relations through buying goods. Love is never far away here – be that in demonstrating parental love or the care of a grandparent towards a child through the purchase of treats, or buying certain (less healthy) foods (and not others) for a partner or children because they know they happen to like them.[37] Then there has been another body of more politically inflected work that has examined the emergence of the identity of 'the consumer' as a political category. It has highlighted the growth in consumer rights and in consumer legislation, and the rise of the idea of the consumer citizen, whose responsibility to society is to purchase.[38] Much of this work takes the historical long view and therefore tends to downplay the connection of the meta-category of 'the consumer' to neoliberalism and globalization. But globalization has undoubtedly propelled the normalization and normativity of consumption in the countries of the global North, where rights to live in particular ways and to consume at particular (high) levels have become not just normalized but actively promoted, encouraged politically and made economically accessible through easy access to credit (or debt-financed money). Think, for example, of how Europeans' holidays and travel have been transformed from the once-a-year trip that was taken if it could be afforded, to be seen as entitlements funded by loans if necessary, and then to rights to multiple trips each year, most of which involve flying – 'winter sun', skiing, the big summer holiday, mini-breaks. Purchasing then is heavily mediated by finance capital, and is required for the continued accumulation of wealth. Hence the response of the G7 countries to the global financial crisis and then to the COVID-19 pandemic, both of which showed the imperative for governments to keep consumers consuming. The conditions of possibility for consumption then are shaped politically; what may look like consumerism is actually the effect of a financialized political economy in which we are all consumers now.

Not surprisingly in an area of research that has been dominated by acquisition and the emergence of the identity of the consumer, rather less attention has been paid to the effects of all this acquisition, or how people actually live with things. A notable exception to that generalization is the body of research that understands consumption as a practice.[39] The turn to practice in the social sciences is very much bound up with recognition that the human world is not defined exclusively by the representational domain – or, meaning and signification. It goes beyond a world that is thought of in terms of object–subject, subjectivities and even identities to encompass routine doings, the kind of activities that we just know how to do without giving whatever it is any real thought – our routine behaviours and activities. Typically, those sorts of activities entail working with our bodies, and with allied objects, but doing so in recognizably similar ways. So, there are often rules, or routines, involved in practices, even if they are mostly tacit. We can think of these as normalized ways of doing certain things. Think of something as banal as making a cup of tea, which, for a tea drinker like me, would be a sequence of (dash of skimmed) milk into mug, add tea bag, add boiling water, whereas made by a non-tea drinker at a Costa or Starbucks it might be lots of (at best semi-skimmed) milk with hot water, with the tea bag passed to you – not the same drink at all! That's an example of what happens when someone doesn't have the know-how or knowledge of the tacit rules and/or doesn't have access to things that are necessary (boiling hot water in this instance, which requires a kettle). Tea-making here most definitely is not 'hanging together', with a big rupture point opening up between the person whose habit is drinking tea and the tea maker's inadequate offer in the name of a tea.

Consumption, in this body of academic work, is not only a practice; it comprises multiple other practices, and it's to those other practices that the acquisition of goods and services relate, and then they in turn connect to other practices in what is an interconnected web. Take clothes shopping for example. This is often portrayed in the mainstream and popular media as an example of extreme overconsumption, and when we look at expenditure patterns and at the wardrobes of celebrities it is easy to see why that is. But, clothes shopping connects to a whole host of other areas of life: to work (where there are very definite standards and norms of acceptable dress), to routine leisure activities (jogging or running, swimming or a host of other 'active' activities, which ideally require particular sorts of clothing to enjoy doing) and to special occasions (where again there are a host of dress codes in place). All of these areas of our lives, then, have sets of sometimes written, but mostly tacit, rules about what (and what not) to wear and, for most people, getting this wrong is not an option. Try being taken seriously at an interview, for instance, if you turn up in a pair of jeans and a t-shirt. Maybe OK in a big tech setting, or in the creative industries, but probably not a good choice in many other settings. Hence why people go out and buy clothes for an important interview; they recognize the need to dress to impress, or at least to show their acknowledgement of the norms of social respectability. Seen from this perspective, then, the amount of clothes in our wardrobes is not really a marker of overconsumption; rather, it's because our lives are so complex that we end up needing this kind of range of clothing – the range helps us to navigate that complexity. And clothes don't just stop with the wearing. They in turn connect to a whole host of other practices of clothing care and maintenance – think of washing, drying, airing, ironing, dry-cleaning. And again, each of those practices needs know-how in how to do them, otherwise one ends up with thoroughly wrecked clothes. Think of whites that end up dull grey or worse because they went into the washing machine with colours, or adult garments that have suddenly become suitable only for a 6-year-old because the washing temperature was too high!

Thinking in this way about consumption has transformed our understanding of how people relate to the things in their daily lives. It has also highlighted that things are not passive, inanimate participants in routine activities and behaviours. Rather, they play a key part in our capacities to do certain things, and in the development of know-how. Think, for example, of the inadequacies of standard everyday footwear in snow conditions: here, the taken-for-granted activity of walking becomes not only difficult but if you try doing that on hills and mountains, the ingress of frozen water, particularly over a long period, becomes potentially hypothermia inducing and therefore dangerous. Serious outdoor walkers therefore have special boots for winter conditions – or, in other cultures, they might resort to snowshoes or skis. And then there are the things in themselves, or what is referred to as 'thing agency'. This means that the capacities and qualities of things affect the performance of a practice. At one level, we see this when people with highly developed competences in relation to a certain activity achieve more with certain things than others with less or no skill. If I tried to juggle balls, for example, I would be totally hopeless at that, whereas a skilled exponent of juggling would be able to work with those balls in ways I couldn't begin to. At another level though, 'thing agency' is about the physicality of stuff. Like everything else, over time the stuff that we live with and through deteriorates in its capacities and decays. 'New' is only ever a transient moment, defined and bounded by the act of purchase or acquisition. Instead, the reality of living with things is that in using them to do various activities, we use them up – some things faster than others. This takes us to the second premise.

Premise 2

If consumption is a practice comprising multiple practices, and the role of things is to enable the routine enactment of practices, then it follows that consumption will also involve the valuation of those things and that valuation is part of the meta-practice of consumption. Some of this will be explicit: how fit are they for purpose (whatever that might be)? Some of it is the type of awareness that develops in the enactment of a practice: that a child is growing out of a bike, for example. Consumption will also inevitably involve the failure of things because, let's face it, things break, they get damaged, they fail to work as intended or they just decay through use, and while some things acquire an aura (or 'patina' as it is termed), others most definitely do not. The effect of all this is that the practice of consumption is not just about the acquisition of stuff; it's also about getting rid of things, or what I term the practice of ridding. How might we think about this?

One of the most influential ways of addressing this has been to focus on discard, which is frequently equated with the term 'rubbish' or 'trash' – the stuff that we throw out, just as our ancestors did. Rubbish leads to the idea of 'rubbish value' proposed by Michael Thompson.[40] This is the third of three categories of value he identifies: transient ('here today, gone tomorrow'), durable (the stuff that lasts and that is often a joy to have around) and rubbish (that deemed of zero value), which is often equated with waste. Rubbish is not a fixed attribute. Things can be reclaimed from the category 'rubbish', often because they have become seen as collectable or desirable artefacts, or even cultural heritage. Working as a carpenter in gentrifying North London, Thompson witnessed 'gentrifiers' raiding junk shops and secondhand shops, seeking out discarded doorknobs and door handles, fireplaces, and fire surrounds, fixtures and fittings to furnish their homes.

Rubbish value itself, as a principle, is relatively simply stated. It applies to all fabricated things, but it has a complex relationship to time that connects with the conditions of fabrication. There are some fabricated things whose value is durable, that is, it stays fairly constant or even accrues with time. Most obviously, these are one-off, unique things. A piece of art such as a Rothko painting or a Stradivarius violin would be classic examples. If the cultural value of the artist as an artistic figure increases, then the value of their art or craft work increases too, provided whatever the thing is has evidenced provenance. Slightly different is the manufactured limited-production run. An early example is Wedgwood, the English ceramics manufacturer, who started to produce earthenware and stoneware in the 18th century.[41] Limited-production runs manufactured scarcity. In other words, they produced what for subsequent generations became the collectable, and a practice of collecting (yet another consumption practice) that has continued into the contemporary period. Limited prints of artworks, or editions in the case of books and stamps, all utilize scarcity to heighten and retain the value of a particular thing or object. It being a rarity constitutes its value for many collectors.[42]

Most fabricated things are not like this, however. They are in the transient category, where value declines over time, in some cases rapidly.

This is especially true for manufactured, mass-produced goods – think of cars, clothing or mobile phones. Their decline in value is a matter of the intersection of consumption (how used something is in everyday life and hence its condition) with fashion and the pace of technological change. A good example would be old cars with very high mileages. Anyone who has tried to part exchange an old vehicle with a car dealer will have encountered the ultra-low price that is offered for such models. Factor in fashion and technological change and whole categories of goods can become of rubbish value. A good example here is devices for listening to music on the move. Whereas the Sony Walkman was the cool device of the 1990s for listening to music on the move, the advent of first the iPod and digital downloads, and then of streaming services, combined with the ubiquity of the mobile phone as the technological interface for large swathes of human life, has rendered it of limited value. The 2000s saw Sony Walkmans being increasingly consigned to the household bin. They had become obsolete – rubbish goods of rubbish value.[43] Then there is the category of mundane, everyday goods fabricated and marketed as disposable or designed to enable other goods to be moved around, stuff like disposable wipes and tissues, cotton wool buds, plastic bottles, bags and plastic packaging, or, to use two examples from the COVID-19 pandemic, lateral flow test kits and face masks. These types of single-use goods epitomize manufactured transience, for it is the immediacy of their very use that often turns them to rubbish value.[44] Then there is a subset of these goods whose very nature suggests disposability – think of the design of a disposable nappy, or sanitary products. All of these things have what are termed 'material affordances' that encourage their being thrown away; their very use ensures that they are folded or compressed in the hand and then let go, discarded – typically in a bin.

While there is much that is insightful in the idea of rubbish value, there are also limits. Two are particularly significant. The first comes from thinking about things exclusively in terms of categories of value; the second is the idea of rubbish value itself and its application to objects. Let's take each in turn.

Thinking through Thompson's three categories of value is another version of the problem that we met earlier. There is a tendency to overlook the process that connects to these categories: valuation. For an object to move in and out of these categories, or through them, requires an act of valuation. So, if we are to understand how stuff ends up in the category that we call 'rubbish' or 'trash' then we need to think not just in terms of the category of rubbish value, but also in terms of processes of valuation. Thinking in this way has turned attention to thinking about ridding as a practice structured by valuation. It highlights that there are different but mostly routinized ways by which people get rid of things. In very broad terms, these involve three choices: giving away, selling or discarding. As we will see, those choices are enacted in routinized ways, hence seeing them as practices.[45] They also correspond less to categories of value and rather more, as I will show in Chapter 4, to three distinctive value regimes that connect respectively to the redistributive economy, to secondhand markets, and to various facets of the waste management industry, and thus to the mainstream capitalist economy. But it is individual acts of valuation by consumers that determine what ends up in which regime and in which part of the market for discarded goods. Valuation then precedes value, and it is always subjective; it is thoroughly dependent upon the person who is making the valuation assessment. It is also therefore always geographically and historically contingent. This brings me to the idea of rubbish value itself.

The first point to make about this is that in Thompson's formulation, it applies to objects (that is, to things with a distinctive form and function). Now this is a problem, not least because objects are not just defined by their capacities to enable us to do certain activities or tasks; rather, they are fabricated things, or, to spell it out, they are made from stuff. This means that there is a latent material value in all fabricated things, in which case, how is it that things can become of rubbish value, or of zero value? The answer is, only if there is a category that renders them exactly that – as of zero value. Now, at the time when Thompson was developing his arguments, that is precisely the situation that existed in the UK, and many other countries in the global North. Things and stuff valued as rubbish were handled as if they were of zero value. They were even talked about as proper nouns: as 'the rubbish' or 'the trash'. So they were seen as rubbish or trash and discarded in rubbish and trash bins. Now things are very different. The phrase 'the rubbish' has pretty much disappeared from the European context, replaced by tighter descriptors such as 'the recycling' and looser terms such as 'the grey bin' – or whatever colour code is assigned to bins reserved for undifferentiated household discard. This tells us that the latent material value in household discard is being recognized by consumers and that use of the term 'rubbish', while not quite redundant, is surrounded by sufficient uncertainty to render it problematic to use. So, even at the site of consumer discarding, the concept of rubbish value is now of questionable merit. But was this rubbish or trash ever really of zero value? Even back then in the late 1970s, I would argue it wasn't, for even then, well before societies started to recognize that they had a waste problem, the waste management business was being paid large amounts of money to collect this 'rubbish' or 'trash', to transport it and dump it, bury it or burn it. Rubbish, or trash, then had value, maybe not to those who threw it out, but most definitely to those who collected it. A second point follows on from this: stuff that is declared to be of zero value is more accurately thought of as stuff that is out of place. Zero value is not absolute. It can only ever be a context-specific valuation. What this also tells us is that for that object or stuff to be revalorized, it is going to need moving to another place.

Premise 3

Since the industrial revolution, the amount of stuff left behind as a result of economic activity has been growing. That has been a result of the expansion in productive activities – or agriculture, extractive, manufacturing-based and service-related activities – and the expansion of consumption. On the production side, this stuff is the residue left behind by processes designed to produce a more valued and valuable something else. It's stuff like the straw stubble left over from harvesting grains; pips; skins and peelings from processing agricultural crops; the mine tailings left over by ore extraction and treatment; the steel slag that is left over from steel production via a blast furnace; or, to use a previous example, hospitality food waste – the stuff that was left uneaten on the plate or from food preparation. This has had profound effects when it comes to how we think about economic activity, not least because this kind of stuff is the stuff that no one is really that interested in, be that farmers, manufacturers and industrialists, small businesses like restaurants and hotels or, as we will see, those who are in the business of analyzing economic activity.

If we take neoclassical economics, for economists, leftovers, residues and wastes are a classic example of what is termed 'negative externalities'. They show that the social costs of production exceed the private costs, and they are acknowledged to impinge negatively on third parties. The most obvious instance of waste as a negative externality is the actual, physical dumping of wastes and residues by businesses – a practice that characterized first the industrial revolution in the global North and then more recently economic development in the former Soviet bloc, China, India and large swathes of the global South.[46] This is the point at which waste becomes pollution; it is where stuff of zero, or negative, value to its current owner (here a firm) is externalized beyond the physical boundaries of the firm, either by being discharged directly into the open environment (rivers, the sea, the air) or removed some place else to do just that. In both cases, the effects – and costs will be felt by unconnected others. This is often called 'exporting harm'. Another example of how waste works as a negative externality is through the burden of displaced intergenerational social costs. The supreme case here is, of course, GHG emissions and climate change; another would be radioactive wastes, where future generations will bear the burden of managing the radioactivity generated by previous generations, be that through nuclear power generation or nuclear arsenals, or both. For economists, negative externalities are a serious problem. They indicate a state of Pareto inefficiency and a breakdown in the system of private property and property rights that underpin market-based societies, for dumping is a form of trespass. The response, which seeks to restore efficiency, is typically to suggest regulatory interventions that seek to make producers internalize those costs, be that through taxes, trading schemes or behavioural nudges. This shows us that for economists, it's not so much the stuff that's left behind that matters but rather what that stuff stands for or indicates, and most especially how its presence disrupts the foundations of economic theory.

Away from neoclassical economics, other influential lines of economic thought have been just as guilty of disregarding the stuff of waste. In the traditions of political economy, for example, the focus on what is termed the 'productive economy', and in particular, the commodity, has had profound consequences. One has been the privileging of agricultural, extractive and manufacturing activities over service activities; another has been a long-standing indifference to, and often neglect of, consumption. Here consumption gets reduced to and equated with the retail sector, and studied in terms of retail capital and labour. Then, there is the effect of the focus on the commodity, which is that wastes and residues were mostly lost from view. For sure, there were exceptions, such as work in the political ecology tradition that recognized the social costs of waste as an externality and has long highlighted the burdens and environmental injustices of dumping. But an effect of this, I would argue, is that wastes came to be seen as somehow outside of economic activity, as stuff that is dumped – pollution, in other words. An effect of this line of thought has been to think of dumping in relation to a sense of a waste commons.

At one level, the term the 'waste commons' harks back to the classic 'tragedy of the commons' argument.[47] Open dumping is seen to have the same effect as the overexploitation of a common resource. Just as overgrazing leads to environmental degradation, so does dumping. There is, however, another sense of the waste commons that is mobilized, particularly within the environmental justice tradition. This develops the sense of waste being somehow outwith the mainstream economy, for it understands the commons as defined by customary rights and an absence of private property. Much as we might think of customary rights in terms of practices that allow people to appropriate foodstuffs – foraging for berries, leaves or fungi, for instance – so we can apply the same logic to what is discarded. Stuff that is cast out is freely available for others to reappropriate. In such a way, the waste commons is seen as the means to livelihoods for waste scavengers.[48] How visible the waste commons is depends on where one is in the world, levels of economic development allied with the degree to which land is parcelled up into private property, and the extent to which alternative economies exist in the interstices of mainstream activity. In the global South, waste scavenging is a widespread, highly visible practice, one that goes hand in hand with uncontrolled dumping. By contrast, one often has to look harder to see it in the urbanized global North, but it does exist – in neighbourhoods where the habit of placing out on the street stuff that is no longer wanted still persists, and in the activity of bin diving.[49]

By contrast, within core traditions of political economy there has been a general unwillingness to recognize the waste sector as a key part of economies. We can trace this lack of interest back to a widespread wariness of non–commodity-producing service activities and to the positioning of waste as the service at the end of the road in what is now labelled as the 'linear', or 'take-make-dispose', economy. If all that the waste industry was about was disposal, then what could possibly be of economic interest here? The short answer is 'a lot', as I now show.

To begin to unpack this we need to note first that the waste industry is not and never has been about disposal in the literal sense of the term. While it might at times have marketed itself as such to those who wanted to get rid of certain sorts of stuff, what 'disposal' means in practice exemplifies how waste moves from being material categorized as of zero value to being revalorized. It entails the collection and removal of whatever the waste is from a particular location and its relocation someplace else. There it is transformed into another material state by a form of processing, which mostly, up until very recently, has involved burning, or incineration, but can also be a mechanical, chemical or biological transformation; either that or it is deposited in what we now term 'landfills'.

As the contemporary excavations of prehistoric middens show, depositing waste, or burying it, is a form of long-term storage mediated by processes of decomposition and decay. So, the residues and wastes of the last 200-odd years of industrial capitalism that have been sequestered in landfills are going to be around for a while yet, for future generations of archaeologists to excavate.[50] While they are a means to long-term storage, landfills are also economic entities; they are very much facilities in the contemporary present of economies. There are two dimensions to this that need to be highlighted. First, landfills are an example of what we might term the 'economization of time and space', or 'making markets out of time and space'. In more straightforward terms, these markets create forms of economic storage.[51] Some of the clearest examples of this come from the kinds of consumer storage facilities that are now pretty much ubiquitous in most urban areas in the global North, and which can be hired by literally anyone. In economic terms, landfill has some similarities. It's a means of attending to overflow. The only difference from the kind of local storage facility available to you and me is that this is a one-off fee payment for storage in perpetuity, for currently what is buried in a landfill is not retrievable (give or take archaeological excavation). That may change in the future, given debates about landfills as potential urban mines.

Second, landfill is the simplest economic form of waste management. As with other primary activities (agriculture and the extractive industries), the chief capital input for landfill is land. The only other costs are operating costs and, at least until recently in the global North, those have been low, as landfill operations require little labour and are highly mechanized. So landfill isn't just simple; all things being equal, it's the cheapest option for managing waste. This is why, the world over, deposition has been the primary means of waste management. The regulatory requirement in certain parts of the world to capture landfill gas (methane), and to export that to the gas grid, has added to those costs, but landfill costs for operators still remain significantly lower than the highly capital-intensive waste management alternative, which is incineration. Modern incinerators are technologically complex, manufactured, industrial-scale processing plants. Albeit they don't require large amounts of labour to operate, as highly capital-intensive plants they will always be more expensive than landfill. These, then, are the absolute economic basics of the global waste management industry and, self-evidently, they feed into the prices that are charged by the industry for the two disposal routes. They also show, incontrovertibly, that for the waste management industry, material categorized as waste is most definitely not stuff of zero value. Rather, it's a source of revenue, while different waste management technologies offer different rates of return and hence levels of profitability. Far from being beyond economies, the shape and form of the waste management sector is better thought of as capital's solution to the problem of waste.

In the mid-late 1990s, waste and particularly landfill became the bête noire not just of environmentalists but of global and international regulators. The catalysts here were two-fold: 1) the success of environmental campaigners who had long sought not just to get wastes recognized as negative externalities but to internalize them, through the twin principles of 'the polluter pays' and managing wastes at or near to their site of generation, rather than exporting them someplace else; and 2) the political recognition of climate science, in particular the role of GHGs in global warming. Methane's potency as a GHG[52] meant that landfill, as a generator of methane, became an obvious target for policy intervention. This occurred first in the EU and then more widely in the global North. A series of interventions, rooted in neoclassical economics, were introduced to address waste as an externality, simultaneously seeking to divert wastes away from landfill and to make landfill if not an obsolete technology, at least the technology of very last resort. The means to this typically was a form of landfill tax levied on waste generators.

It is no exaggeration to say that these interventions have proved to be the single most disruptive act in waste management in over 100 years, and their force continues to be felt and reverberate globally. That regulatory push has resulted in the transformation of the waste management industry into a waste and resource recovery sector. Underpinning that is the extension of commodity relations into major parts of the sector, for wastes, rather than just being managed, are now treated and then bought, sold and traded as secondary resources in the commodity markets, or, alternatively, converted into (renewable) energy resources: electricity and gas. Waste, then, has become a valuable commodity. Not only does this spell trouble for the notion of the waste commons – for, as with all stuff that becomes of value, it is rapidly turned to private property and enclosed – it also means that waste (and the waste industry) has attracted the interest of finance capital. As I will show in Chapter 6, waste is not just a commodity, it's also an asset, and with that comes projections of the generation of future waste. As I will show, the financialization of waste has meant that waste, and the waste industry, is thoroughly incorporated into the heart of 21st-century capitalism. What has evolved is a capitalist fix to the problem of waste, one that demands the production of more waste, even as it is argued that interventions reduce it.

Outline of the book

Premises 1–3 provide the structure for the remaining chapters.

Chapters 2 and 3 develop Premise 1. In Chapter 2, I show how human social life and therefore consumption is underscored by acts of discarding. Discard is but a manifestation of this. Discarding is argued to be one of the means by which people create order, both in individual lives and at the level of societies. Once this is recognized, it is clear that discard is a foundational aspect of all human life; it cannot and will not be eliminated. However, there is no inevitability about the scale of that discard and that discard will become waste, which is a societal and political choice. In Chapter 3, the focus turns to consumption and the central role that this plays in the economies of the global North, China and a growing number of middle-income countries. For many countries, consumption is fundamental to continued economic growth, so the acquisition and replacement of goods is baked in to many economies, most notably through planned obsolescence. An inevitable consequence is the discarding of goods.

Premise 2 is the focus for Chapter 4. Here the emphasis turns to how consumers discard stuff. At the heart of this are three different value regimes and how people assign things to them through processes of valuation. The regimes are hierarchically related by the apparent value of goods in secondhand markets. So, goods can and do trickle down through these regimes, although where something is placed initially is totally dependent upon the initial valuation accorded something by the person doing the divesting. At the same time, value regimes overlap and compete, chiefly through convenience. It is here that the conduits that connect discard to the waste industry have a competitive advantage, precisely because they offer consumers easy routes for discarding stuff. Almost all other routes – be they giving stuff away or trying to make money from them, and be they digitally mediated or not – involve consumers in considerable amounts of work. They not only require caring about discard but making time and space for the practice of discarding among all the other practices of consumption. This is not only in short supply, it's finite, which means that discarded goods cascade to the waste industry.

Chapters 5 and 6 elaborate on Premise 3. They show how waste is thoroughly part and parcel of the global economy and how the waste industry has been embedded in 21st-century capitalist economies, through commodification and financialization. In Chapter 5, the focus is on the recommodification of wastes in historical and global perspective. This is the stuff of the recycling industry. Taking the historical long view shows the importance of economic conjunctures if markets in goods containing a large proportion of recycled content are to prosper. It also highlights the challenges such goods face from primary manufacturers. Those challenges have not gone away. Indeed, as I show, the core challenge facing innovation-centred solutions to 'problem wastes' that seek to turn them into secondary resources is not so much the technical one – of turning them into something that can be used – but economic. In market-based societies, no amount of technical wizardry is going to work if what results is a poorer-quality, higher-priced substitute for a dominant product in the market. That is the harsh economic reality that faces much resource recovery activity in the global North. It is also one of the reasons why the global trade in wastes persists, and why the recommodification of wastes remains an activity concentrated largely in Asia and other parts of the global South.

In Chapter 6, the emphasis turns to the discard that actually becomes waste, or residual waste as it is termed by the waste industry. Again, the approach is to take a historical and global perspective. The chapter charts the development of largely nationally bounded waste regimes and the early competition between the technologies of landfill and incineration, before documenting how, for much of the world, landfill became 20th-century capitalism's primary solution to managing wastes. It then shows how, beginning in the 1990s, this waste regime has transitioned to a waste-to-resource regime and within that how landfill has been supplanted by incineration as the means to manage the world's residual wastes. The global move to incineration has opened up major new markets for the waste industry, creating transnationals comparable in scale and scope to medium-sized car manufacturers. It has also created huge new markets for the firms who design and manufacture these plants and their core components, chiefly firms headquartered in the global North. Even more important is that it has allowed waste to be captured by finance capital. I close the chapter by elaborating on the financialization of consumer waste and its implications, showing that consumer waste has become a means to wealth generation and accumulation.

In all of these chapters, I highlight the implications of the arguments made for current waste policy. By framing the problem as one of (too much) waste, interventions have sought to reduce waste through the promotion of repair (Chapter 3), reuse (Chapter 4) and recycling (Chapter 5). When we frame waste as an effect of the relationship between economies and consumption, the difficulties with these interventions become apparent, and I make those explicit. I conclude in Chapter 7 by formulating a different set of messages grounded in the relationship between economies, consumption and waste.

2

Discard, Social Order and Social Life

Or, discard is foundational to understanding waste

Although waste is often described as that which is discarded, the relationship between discard and waste is complex and incommensurate. What is discarded does not inexorably become waste. Conversely, much of what is discarded does become waste, if not directly then certainly indirectly. Yet discard is prior to waste – for something, anything, to become waste it must first be discarded. So, to understand waste, we first need to understand discard, not simply as the stuff that is discarded but, more importantly, as an act of discarding. The act of discarding is psychological, social, cultural, economic and material, and always spatial and geographical. But before we can explore these dimensions and their relationship to discard and discarding, there is a need to step back further.

I want to begin this chapter with what might seem at first sight a provocative statement, especially for a book that is about waste. This is that discarding is hard-wired into us humans, just as it is for many species. Discarding is what we do; it's part of what makes us human and what it is to be human. This needs to be recognized and acknowledged, especially when devising interventions that seek to minimize or reduce waste. But unlike other species – think of birds separating husks from seed as they eat and dropping them to the ground, or cleaning up their nests by picking out their guano and dropping it overboard – human discard is qualitatively different from that of other species. It is more than organic. So, it transcends the discards of those material fundamentals of life: food, edibility, eating and its material consequence, shit. Indeed, human discard can comprise literally anything that humans live with and around. In the academic field of discard studies[1], discard even encompasses humans themselves, in the form of discarded places and discarded people, or those places and people abandoned and/or left behind, sometimes as a result of political unrest, conflict and turmoil, but often as a result of transformations in economies. Indeed, the field of discard studies is so all-encompassing that its orbit includes not only this but also much of archaeology, in the form of abandoned settlements and buildings, as in ruins.

My concern in this book is not with this expansive sense of discard but rather with discard in its narrower sense, with the kind of discard that mostly becomes waste and that ultimately feeds the waste industry. This kind of human discard, as stuff, is materially diverse, and in line with scientific and technical innovation, it has got more and more complicated with time. That complexity, together with the inorganic character of much of this discard, means that human discard is more likely to become waste than the discard of other species. This is for solid ecological reasons. In ecological terms, the material character of much of human discard is such that there is no capacity for it to become the basis for the mutually symbiotic relationships that are to be found in the natural world. There, what is abandoned, leftover and rejected by some species becomes the food (or nest material) for other species. A lot of what is discarded by humans, however, is left mostly, if not exclusively, alone by other species, untouched. There are exceptions where this generalization breaks down, such as discarded food[2], while other forms of discard, most notably plastics, are ingested directly or indirectly by other species, with the working hypothesis being that they are mistaken for 'food'.[3] There are other examples where other living things become entangled in the pollution emanating from human activity, notably oil spills, but the general principle holds. Rather than being scavenged, or eaten, much of what human beings discard, most particularly inorganic manufactured stuff, is disregarded by other life forms. As such, it becomes abandoned stuff, of apparently little utility or value. This is what we humans classify as waste, and it is one of the things that makes humans unique in ecological terms.

The realization of that ecological distinctiveness – that our capacity to make waste is one of the defining characteristics of human societies – has led to at least two interventions that have sought to learn from the absence of waste in the natural world. They seek to recast human relationships to the material world through ecological principles. One of these is industrial ecology, which emerged as a field in the late 1980s. Taking its cues from ecology, and in particular the role of scavengers and decomposers in ecosystems, it argued that to minimize waste, economic systems needed to maximize reuse and recycling.[4] The second, and more recent, is the endeavour to recast economies through more circular principles. The antecedents of that are typically identified with Ellen MacArthur, the round-the-world yachtswoman, whose observations on the amount of waste, and particularly plastic, afloat on the oceans led her to create the foundation in her name, which seeks to promote a circular economy.[5] Later chapters will discuss both in greater depth, but for the purposes of this chapter, I want to stick with discard and more particularly what's going on when humans discard things.

The chapter begins by looking more closely at how discarding is hardwired into humans, by examining it as an embodied act, at times cognitive, at others more habitual, subconscious even. Having established these foundational universals, the chapter moves on to show how discarding relates to the dynamic process of ordering social life. Order and ordering operate at the level of individuals, through their relationship to a sense of self and the life course, and at the level of society. The principles that connect discard and discarding to social life and social order hold across multiple societies and cultures, so the examples I draw on span cultures. I close the chapter by arguing that the ubiquity of discard and discarding has implications for interventions that seek to reduce waste.

Discard, disorder and displacement – and their connection to the human hand

For as long as humans have devoted time and energy to thinking about and endeavouring to understand our world, we have also examined what sets us apart as a species. This is the kernel to evolutionary biology. While much of the early literature in human evolutionary biology concentrated on the human brain and mind, another strand of work focused on the differences between the human and primate hand. The capacities of the human hand are identified in this literature as foundational to toolmaking and thus to the emergence and development of technology.[6] That uniquely human capacity to explore and work with the material world, to order it and fabricate from it things that are then utilized in, and in turn shape, human societies is what has really set us apart as a species, to the point that the effect of the human-made is now debated as a unique geological epoch: the Anthropocene.[7]

But let us focus on that hand. Recently, evolutionary biologists have noted morphological differences between human and primate hands. They have emphasized the capacity of the human hand to grip with precision, to rotate and manipulate, to utilize fingers independently and in opposition with the thumb.[8] Those capacities give humans immense advantages over other species – they allowed for the making and manipulation of early stone-age tools. They were also critical to the later development of writing, through the capacity to hold and manipulate a stylus, quill and then pen. It is not too much of an exaggeration to say that the human hand's capacities, in conjunction with the human brain, underpin the development and subsequent transmission of all forms of human knowledge. Yet other professions rely on the hand's creative capacities. Pianists and string players, for example, all talk about 'working hands', or hands that need to work well for them to continue to play an instrument.

And then there are the caring and compassionate professions – healthcare and social care – where the hand's capacity to touch and to offer empathy and feeling is paramount. All these are what we might term examples of the generative capacities of the human hand. But, at the same time, the hand has the capacity to relax, to let go of something, be that another hand or another object, and to do that purposefully or unintentionally, as when we drop things. It also has the ability to chuck, or throw, with direction and precision. Think of how a cricketer crafts how they bowl a ball: speed, spin, curvature and flight are all coordinated by how the hand and fingers work in tandem with the arm as a unit. At the most elemental level of bodily capacities, the relaxation reflex, along with the capacity of the hand/arm combination to throw things away from us, to make them distant from us, underpins the act of discarding by humans. Discarding, as an embodied act, involves letting go; more forcefully, it is about jettisoning, or expulsion. As such, it is always spatial; it involves separation and displacement from the human body. To begin to elaborate on this, here are two examples of the type that often find favour with philosophers. They are small, seemingly insignificant, almost trivial in nature, but precisely because of this they speak to a wider truth.

The first example is a personal tale. I am walking along through the local woods on one of my standard daily walks. It is winter and it's cold and wet. My nose, as usual, is producing copious quantities of 'gunk' so I'm blowing it periodically into a paper (disposable) hankie, or tissue, which I stuff in my pocket as I walk. As I come back towards home, retracing my steps, I notice a paper hankie lying discarded on the path. Was that mine, or is it someone else's? Of course, I have no idea. But, after a minor amount of internal wrangling with myself, I decide that I cannot leave it there littering the path. It might just be mine and it just won't do to be the person responsible for discarding this kind of stuff into this habitat. So, I gather a few more hankies from my pocket, bend down and wrap them round the discarded one and carry them home, at arm's length, to chuck them straight in the household grey bin, reserved for anything that can't be recycled, before proceeding to wash my hands thoroughly.

The second example is a tale told to me by a family acquaintance who, on hearing about a research project I was then conducting, declared herself to be a 'hoarder' before proceeding to give as an example her picking up of rubber bands that she saw routinely discarded in the street by her gate. The telling of this story was elaborated with a string of wider mores – of those who litter, and those who don't, and of those who clear up after others. But, as was pointed out in this exchange, when it was revealed that these rubber bands were always in the same place, there was probably an underlying reason for such a consistent pattern of discard. And so it turned out. Later it was relayed to me that the rubber bands were an effect of the post delivery system. Bundles of mail for neighbouring houses were grouped together for efficiency of carriage and distribution by the postie, who was then left with the problem of what to do with the rubber bands. Evidently, the answer was to discard them en route.

The two examples work as a pair, with similarities but also differences, which allow for establishing a whole host of points about discard, its relation to the hand and its connections to waste.

Let's begin with the similarities. At first sight, both examples seem to illustrate what the anthropologist Mary Douglas famously described as 'matter out of place'.[9] Douglas' arguments are often portrayed as the bedrock of waste research and scholarship. But, as is often the case with bedrock, they've taken on something of the status of the myth, with the result that the phrase is invoked without too much thought and we all move on – 'nothing to see here'.[10] Pause a while, and there is. As Douglas explicitly stated, trash, rubbish and waste are not matter out of place, precisely because they have a place: the bin, or in today's world, a number of bins. Instead, she reserves the phrase 'matter out of place' for a different category: dirt.[11] For Douglas, thinking about dirt is always connected with systems of classification; to make reference to 'matter out of place' implies both a set of ordered relations and contraventions of that order, which reject that order. Correspondingly, dirt – as that which contradicts order – has to be eradicated. Dirt, then, is always about power. The examples in play here, then, are not matter out of place in that sense; rather, they are both instances of the category 'litter'.

Litter is one of the best examples of a different sense of the phrase 'matter out of place' – one that relates to space. When litter occurs, as in the instances of the paper hankie and the rubber bands, it is not where it's supposed to be spatially. As a term, and literally, litter, then, defines the state 'matter out of place'. But does that make it 'dirt'? Not necessarily because when we think about litter, its manifestation is rather more an effect of systems than either a contradiction or a rejection of them. Max Liboiron, for example, argues that anti-littering campaigns such as Keep America Beautiful work to actively maintain the existing order of industrial production.[12] They do this, he maintains, by deflecting attention away from the manufacture of the disposable goods that constitute most litter, like paper hankies, to individual personal responsibilities not to litter. Nonetheless, litter has to be put back in its rightful place for systemic order to resume. It can't stay as matter out of place because otherwise it would act as a disruptive force. Litter, then, commands attention; it demands action. At the systemic level, vast amounts of resource are committed to the reordering of litter. One only has to think of the sums committed by municipalities to street cleaning and cleansing operations, or clearing up public parks, to see that; or, rather differently, of the emergence of mass participation litter-picking as an environmentally motivated intervention to address the rise in beach pollution. Sticking with the paper hankie and the rubber bands, though, allows for drawing a different line of connection to Douglas' arguments.

At the simplest level, the paper hankie and the rubber bands are actionable because they come to notice.[13] But what is actionable varies according to individuals. Let's be honest: would I have picked up those rubber bands? The answer is, if they were right outside the house, possibly, but not necessarily. Would I have felt any moral imperative to pick them up? No. It's a very different state of affairs with the hankie, where it's the very possibility that it is me who has done the littering that shapes the action. But, and this is the crucial bit, if the hankie had not been where I'd previously walked, would I have had the same reaction? No. Rather, I'd have left it alone, untouched. This is because of how the discarded paper hankie intersects with pollution taboos.

Engaging with litter requires that people handle it, directly or indirectly, that they become proximate with it. This is where hands really start to matter. The paper hankie may not be 'dirt' in Douglas' sense of the term but it is an example of a symbolically 'dirty', threatening, risky or contaminating thing. Obviously used in appearance, it is recognized to bear the traces of bodily fluids – possibly mine but, much more dangerously, possibly someone else's. These are not purifying or cleansing bodily fluids (Douglas uses the example of tears to illustrate that state), but rather a more dangerous, potentially contagious one: mucus. Hence the complicated rigmarole that we see played out of gathering together a wad of other hankies to surround this one in order for the hand to pick it up safely. Significantly, this kind of protective gathering and picking up also takes the form of taken-for-granted, or habitual, action. It is unthinking – the type of action that tells a lot about underlying cultural norms. Then there is the holding at arm's length – a subconscious distancing of self from this contaminating other stuff. After which, note, hands are washed – a symbolic as well as physical cleansing indicative of norms around public health and hygiene. By contrast, the rubber bands seem to convey no such sense of threat. Instead, it is their perceived utility that underpins their salvage, along with the capacity of the manner of their salvage to narrate a particular sense of self. Here, touch and the work of the hand is anything but socially risky; it's socially affirming. It therefore takes place entirely unmediated and free of any cleansing, purification rituals, such as handwashing.

Finally, there is the manner in which discard as litter connects to waste in these two examples, or not. This is another point at which the examples diverge, for while the rubber bands are salvaged, the hankies are not. Instead, they are placed in a bin that connects them directly to the waste stream, for that is the placement that is deemed, unthinkingly, appropriate for them. This litter is waste and it belongs here, in the bin. So, the example works to show that the bin is the means being used by this society to order waste. More than this, the original discarded hankie has gathered more materials to it and rendered those clean, unused things waste as well. In such a way, it can be seen that things regarded as socially 'dirty' have the capacity to contaminate that which comes into direct contact with them. This has implications, most especially in relation to the humans who work with discard and waste.

A brief note on 'dirty' work

The connection between dirt, waste and social contamination is engrained in the organization of how human societies manage waste and who gets to work with it. As Jo Beall remarks, 'all over the world waste workers are stigmatized and are likely to be from marginalized groups such as ethnic or religious minorities or rural migrants'.[14]

In traditional societies, or societies where religion is the primary means to societal organization, these connections often show the workings of social stratification. In the most starkly stratified societies, sharp, hierarchical distinctions are made between social groups. India's caste system, notwithstanding modernizing reforms, remains perhaps the most widely known example of such a system, and in that ancient Hindu religious system, the work of dealing with dirt and waste was assigned to those outside the caste system. A raft of current South Asian scholarship demonstrates that caste remains a category through which social stratification and exclusion persist, with waste and recycling work continuing to be associated with those whose ancestral roots are in particular castes.[15]

If caste provides the starkest example of how systems of social stratification order who does the 'dirty' work, there is a plethora of further examples of how, in other societies, ethnic and/or religious minority difference is mobilized to the same effect. In Egypt, for example, at the turn of the 20th century, migrants from desert oases in Upper Egypt started to collect waste from Cairo's wealthy households, later subcontracting parts of this work to the Zabaleen, a Coptic Christian group whose pigs provided an outlet for organic wastes.[16] Slightly different, but nonetheless illustrative of the same principle by which societal outsiders or outcasts come to be associated with waste, are the European Roma, a travelling people, many of whom have livelihoods grounded in waste picking, scavenging and trading in secondhand goods. Recent scholarship in Central and Eastern Europe has highlighted how the associations between Roma and waste are intensifying. Amplified by right-wing populist regimes, these associations are producing a politics of exclusion founded in environmental racism.[17]

In those parts of the world where societies are less obviously stratified, who gets to do the 'dirty' work is mediated by the workings of labour markets. Differentiated labour markets have the effect that certain groups of people (notably BAME, women, older people with fewer recognized skills and qualifications, migrants) are more likely to be found doing low-skilled, low-paid work and in particular types of employment than others, notably white men. So, when we look at who is cleaning offices, hospitals or hotels in the

UK, or at who is pushing a street cleaning machine, or at who is working as a waste operative on a local authority collection round, then we see some of these effects. There are a disproportionate number of BAME people doing many of these jobs, particularly cleaning. In the largely invisible world of waste work, the effects are even clearer. In the UK, in recent years that work has been the preserve of Eastern Europeans and international migrants and refugees.[18]

Focusing on who gets to do waste work shows the lines of distinction and division in any society. Those distinctions are at their sharpest in highly stratified societies, where to work with waste and discard is to reinforce the social system that produces the distinctions. In societies associated with capitalist forms of economic organization, the workings of labour markets ensure that 'dirty work' as a category works to reproduce social inequalities. The world over, what this means is that those who do this work are rarely doing it through choice; rather, they often have no choice but to do it. By contrast, those who can avoid it do.

★ ★ ★

So far, the discarded things under consideration in this chapter have been relatively straightforward, simple and singular things. But discard can be, and mostly is, a whole lot more complicated in material terms. It can be more heterogeneous and more voluminous than these examples, and altogether less obvious in its form than a paper hankie and a rubber band in terms of what it is and what it comprises. What does this kind of discard tell us about discard and acts of discarding?

In the next two sections, I again work with two examples, using these to tease out some general principles about discard and discarding. Both examples relate to two of life's material fundamentals: eating and drinking. As geographically universal and recurrent activities, the two examples result in a serious amount of physical discard. As will become clear, though, there is still the same connection of discard to displacement, to order and disorder. More critical is that the two examples go beyond philosophical insights. They help establish that the act of discarding is foundational to ordering human social life, or that acts of discarding do some real sociological 'heavy lifting' work. The two examples show, in turn, how discarding relates to the restoration of what is termed the psychological subject (or, a sense of self) and to the annual rhythms of social and cultural life.

Eating, drinking and discarding 'on the go' – mobility, fast food and auto-culture

In recent decades, one of the key changes in sociality in societies where access to a car is widespread has been in how and where humans eat. Eating is no longer just a meal, prepared and/or purchased to be eaten at home, together or alone; it is also about eating out, away from the home.[19] And eating out is no longer something that is done solely in the environs of a restaurant, bar or café. As societies have become more mobile[20], and living for many has become characterized by more intense movement over what are often greater geographical distances, so eating too has become something done on the move, be that on foot; in the street, as with street vendors in so many cities in South Asia say or takeaway outlets in European and North American cities (think McDonald's, KFC, Starbucks and Costa); or in cars, via the growth of the 'drive-thru'.[21]

There is a lot that is wrong with fast food and drink, for animals as much as humans.[22] But, fast food and drink is also a major contributor to the amount of discard that is currently being generated.[23] This is because it is typically provided in large, oversized portions that are often too much for people to consume, and because it is supplied to consumers in single-use, disposable receptacles – cups, trays, bags. Currently, an estimated 1 trillion disposable cups are produced annually in the world, with an estimated 3 billion of those used in the UK, of which only 4 per cent are currently recycled. In part, this is because disposable cups are problematic to recycle but it is also, in large part, because they spell 'disposability' to consumers and are marketed as such.[24] Disposability, as we will see in the next chapter, is one of the primary reasons for waste generation; it means things are designed to be used once and then discarded, as in jettisoned or thrown away. So, large numbers of people purchasing this kind of food and drink, and frequently, is what lies behind the growing amount of this discard.

Food and drink sold for consumption 'on the go', however, is even more problematic in terms of discard than 'fast food' per se. This is because it is sold to be consumed on the move and that could be literally anywhere – sitting on a bench in a park, on a bus or a train, as well as in a car. That 'anywhere' spells big problems for the waste management industry, whose primary infrastructure for collecting discard is attached to spatially fixed locations, notably dwelling units (houses, flats and apartments) and business premises. In contrast, the discard of human mobility is dispersed and intrinsically unpredictable in its location. That makes for inefficient and therefore costly collection. Even in city centres, fixed-point, publicly accessible bins to dispose of this kind of trash are often few and far between – something that is not unconnected to heightened concerns over the security of cities and the ease of use of bins for terrorist offences. So, in the absence of bins, what tends to happen is that the detritus of this kind of eating and drinking tends to get casually left behind by people. It might not be dropped, inadvertently or casually, in the manner of the hankie and the rubber bands, but it most certainly is rarely carried far by those who consume it, and it is even less likely to be carried back home or back to an office. Look around any city park or walk through a long-distance train arriving at a major railway terminus to see something of this. There are bags of the remains of 'on-the-go' food and drink left on park benches, by bench legs, underneath trees and around overflowing bins; and on the trains, tables, seats and luggage racks feature exactly the same sort of discard.

What happens to this stuff? Mostly, it is collected up, as efficiently as possible, often by teams contracted to clean up the leftovers of our mobile lives. There have been some experiments recently whereby waste management firms have teamed up with other partners, including some of the major coffee chains, in ways that might allow for capturing some of this material for recovery.[25] By and large, though, to achieve efficiency in collection means that most of the discard becomes waste. Working at pace, to clean a long-distance train say, on a turnaround of less than half an hour, or to clean up a park at the end of each day, means that all that has been discarded is quickly amalgamated to produce piles of heterogeneous materials – paper bags meet cups, some still containing liquids (coffee, tea, chocolate, Coke, Pepsi, slush, smoothies), meet half-drunk cans of beer and lager, meet straws, meet multiple examples of half-eaten food. It is the kind of discard that very rapidly becomes a cross-contaminated, material mess. Those characteristics turn it very quickly to the category of waste. So, its destination is most likely either incineration or landfill.

Rather different is the discard that results from the drive-thru meal. Take a look at many of the roadsides in the UK, near to wherever McDonald's or KFC drive-thru outlets are, and you will find that the remainders of these meals and their packaging are scattered everywhere.[26] This is because bags and bags of these remainders are literally jettisoned from car windows. So, pretty much everywhere that cars go, so too do these bags and their contents. But, while the cars (and their passengers) travel on, the bags are left behind, where they land, as litter.

Aside from associating this discard with drive-thru businesses and pushing businesses to 'nudge' their consumers to 'do the right thing' with the leftovers of these meals, much of the attention regarding this phenomenon has focused on it as an example of littering.[27] Typically, this kind of commentary is also accompanied by exhortations to the public to 'take your litter home'. Rather less attention has been paid to why this jettisoning might be occurring. Why do people do this? Are they really, as anti-littering campaigns imply, rampant litterers, seemingly with no care for others and the wider environment? As ever, things are likely a little more complicated.

Such moralizing, I would argue, overlooks two things. The first is the lack of design for trash in the interiors of cars themselves. While car interior design has caught up with some aspects of 'on-the-go' food and drink, notably through recognizing the need for drink holders, it hasn't quite recognized the discard that is left over from in-car food and drink consumption. Unlike homes, cars – as yet – do not have bins attached to them, and drive-thru businesses do not yet supply their customers with a means to collect their trash, other than the packaging that the food is dispensed in. Chucking out the leftovers, then, I would suggest, is a means to consumers keeping their car interiors tidy. But a second oversight is perhaps even more significant. This is the smell that emanates from leftover fast food and its capacity to affect the interior air of a car's ventilation system. This stuff smells disgusting, even just a short while after it has been consumed, and the instinctive reaction of all humans to disgust is to separate themselves from it, as fast as possible. In this instance, that means jettisoning it out of the car window. In short, the olfactory qualities of leftover fast food foreground that 'the what' of discard can often be in the category of disgusting stuff, or the 'yuck factor' as it is sometimes referred to. More broadly, this connects with a body of academic work heavily influenced by psychoanalysis. Here, the act of discarding is entangled with managing a state that psychoanalytical work labels as 'abjection'.

A brief note on abjection

At the level of the individual human psyche, the relationship between discard as stuff and the act of discarding is interwoven with the abject and the state of abjection, respectively. These ideas are most strongly associated with the work of the French feminist theorist Julia Kristeva.[28] Building on the work of the French psychoanalyst Jacques Lacan, whose work highlighted the importance of filth to the formation of the subject, Kristeva argues that the state of abjection occurs when an individual human subject is confronted by their corporeal reality, that is, by their existence as a body. This body is living, breathing, fleshy; it has its own agency and capacities, as well as desires; it is leaky, in the sense that it produces stuff that exceeds body boundaries (excrement, urine, menses, vomit, blood, mucus, tears, ear wax); and it is also a body that will become a corpse, thereby transcending the self. So, the body poses trouble for the subject; a consideration of taboos across multiple societies tells us this. How do societies order and attempt to control these bodily capacities?

When bodies assert, they are frequently managed socially by placing them in discrete, spatially differentiated – or liminal – spaces. Liminal spaces are spaces that exist on the margins and edges, or in the cracks, of social life as it is conventionally ordered. The menses hut of many traditional societies is but one of the most obvious examples, where menstruating women are separated off from others, mostly men, as unclean. The same separation of those people and substances denoted to be polluting or unclean underpins much of the spatial ordering of developed societies. Think of how hospitals corral those who are sick and separate them off from the wider community; of how the mortuary functions as the residing place for the corpse pre its disposal; or of the work of the flush toilet, which uses water (itself symbolically cleansing and purifying) to remove human wastes from the dwelling place, to render them instantly out of sight (and out of mind).

Of course, bodies never go away; they continue to assert their presence, to the point when all that remains is a corpse, which is the point at which we too become subjected to the technical processes of managing disposal, be that through burial or cremation. On a day-to-day basis, too, our bodies have the capacity to assert themselves. Typically, these are moments when body boundaries are transgressed in a social situation. Small examples would include sneezing or coughing in a concert say, a nose bleed or flatulence in a meeting. When they do erupt, such moments entail the breakdown of the key self/other distinction that is fundamental to much social life and to social order. They mark the reassertion of an emotional, instinctual and corporeal world beyond meaning and the return of the abject – or, in psychoanalytical terms, all that is rejected by identity. Confronted by this stuff, the response is frequently of shame and embarrassment, but it can also be of repulsion, disgust, shock and horror even. When this happens, and in those circumstances where this is possible, casting out, or jettisoning, is the means to restoring identity. The act of discarding, therefore, is the means to the return of the subject.

So, let's go back to those jettisoned bags containing the leftovers of 'on the go' food and drink. As well as being vital substances, food and drink are potentially polluting. Our bodies require food and drink just to live, but edibility and ingestion also carry dangers. Food and drink can make us sick – in some cases, very sick indeed. They can poison us, and the effects of that can, if untreated, kill us. Food and drink can also be a vector for the passage of infectious disease. Eating infected species (mammals, fish and birds, typically) can be the means to zoonosis, while drinking contaminated water is one of the primary ways by which bacterial and parasitical infections are passed to humans. As humans, we know this, and we fear this. These fears assert when we travel off the beaten track, where we are likely to encounter all manner of things that are regarded as edible by other cultures but not our own. When confronted by insects, or by animals and parts of animals not regarded as food in their own cultures (examples would include the meat of animals regarded as pets, or eyes and brains), the reaction of many western people is one of repulsion and disgust. This shows that food can be abject. But this is 'other people's food'. What about food that is more obviously 'in culture', like fast food?

On a mundane level, all edibility and ingestion involve orality. The mouth is a key body boundary; it is one of the means by which the lines between cleanliness and defilement, or purity and danger, are lived, by all humans, men as well as women. What that means is that eating and ingestion are inherently open to being seen as troublesome. This is what I would argue is going on with those McDonald's bags. Something that smells that disgusting pretty much as soon as we've eaten it is really troublesome stuff, because that stuff is actually stuff that has been ingested. Literally, it's inside us, working its way through our digestive process. Jettisoning that McDonald's bag is not just a means to keeping car interiors tidy; it's a means to dealing with that disgust. It allows us to restore our sense of who we are, in the face of evidence (physical and olfactory) to the contrary.

★ ★ ★

Discard and acts of discarding are not just central to the conduct and management of the self; they are built into the social order and ordering of all societies. In the next example, the focus is on the annual rhythms of social and cultural life and their marking through festivals, feasting and partying. As will become clear, there are very strong connections between these times of celebration and acts of discarding. Discard is intrinsic to the doing of festivals, while acts of discarding are critical to differentiating festival time from the normal social order, and its resumption.

Discard, discarding and the marking of festivity

Festivals, or periods of societal celebration, are characteristic of all societies. And therein is their connection to religious belief systems. The world's major religions have an annual calendar shaped by ritual in which religious observation is marked on the one hand by periods of piety, abstinence and penance, punctuated and alleviated on the other by times of celebration and festivity. Christmas, Easter, Navratri, Diwali, Eid, Passover and Hanukkah are major examples of these festivals. Notwithstanding its connections to puritanical traditions, Thanksgiving is secular America's equivalent festival – four days of consumption, focused on eating, shopping and ball games.

For those who live lives ordered by faith, the annual rhythms sit alongside a pattern of daily religious observation; together, they give meaning to the world and shape identities. So much so that the relation between fasting and festivity is seen to be deeply interdependent. Can one really do Eid al-Fitr without having observed Ramadan? Can one celebrate Easter without having observed Lent? Those questions may not trouble the more secularly minded, but even so there is still a strong relationality between what is demarcated as festival time and the rest of daily life. This is because festivals mark periods when normal life – notably work and employment, and education – is suspended. Instead of economic life being prioritized, this holiday time valorizes social life and social ties.

In those societies where religion either has not been or is no longer tolerated, there is still a place for festival. In China, Chinese New Year is the equivalent to a secular Christmas. It's a two-week shutdown of all employment and the suspension of routine life. Millions upon millions of Chinese return to their family home for what is known as the Spring Festival, be that from the US and Europe or from the factories of southern China.

Festival, then, transcends different types of meta-societal organization. It is a fundamental characteristic of human societies, something which is foundational to being human and a key means to the annual ordering of societies. The world over, and alongside seen to be necessary acts of religious observance (such as lighting the lights, attending a faith meeting, special prayers or a church service), these major festivals are marked by two primary sorts of human activities: feasting and gift exchange. Both involve large amounts of discard.

Let's take the feasting first. Feasting has long been connected to displays of social power. The 'ruinous feasts' associated with the aristocracy of early modern England, and their attendant practices of the extreme overconsumption and deliberate wasting of food, were a means to demonstrate social standing and authority and were a forerunner of what Veblen termed 'conspicuous consumption'.[29] Classic anthropological texts have identified food being used to similar purpose within traditional societies. Malinowski's account of the Trobriand Islanders, for example, includes gifts of large quantities of yams being deposited outside the recipient's home and left there to rot – testimony to the inability of the recipient to provide similar in return, and to the debt they incurred to the giver as a result.[30] In contemporary societies, we may no longer feast on quite the same scale or deposit piles of food outside one another's homes, but we continue to mark festival days with the feast.[31] Think of the typical traditional Christmas meal: an enormous turkey, mountains of roast potatoes, roast parsnips, Brussels sprouts, carrots, gravy, stuffing, bread sauce and cranberry sauce, followed by Christmas pudding, laced in brandy butter, and multiple mince pies. Then multiply that across UK households. At Christmas 2019, an estimated 10 million turkeys, 25 million Christmas puddings and 175 million mince pies were purchased.[32] It is the same with the major Hindu and Muslim festivals. Multiple media outlets are full of examples of perfect menus for Diwali and Eid, where 'everyone's needs can be met' and 'where there is enough for everyone'.[33] Those phrases translate to mean a large table, stacked to capacity with plates and plates of home-prepared food – at minimum, a choice of two starters, two mains, three accompaniments and a dessert for each meal, plus classic favourite snacks and nibbles. Similarly, Chinese New Year is celebrated with an array of food considered to bring good fortune, wealth, health and happiness in the coming year.[34]

At a time when debate over the politics of food is dominated by issues of obesity, poverty and the capacity (or not) of intensive animal-based agricultural food systems to continue to be able to feed the global population, it is often difficult to look at these kinds of groaning tables without resorting either to moralizing or to a vague sense of disgust, which marks the return of the abject. Seen from the outside, and listed like this, such amounts of food seem, well, obscene, gluttonous, unjustified. But that is just the point – it's a view from the outside in. When we are sitting on the inside, sharing in this sort of exceptional meal, what it is about is just that – a sharing of abundance with significant others. The pleasure lies in that. It's a mutual enjoyment of there being more than enough to pass around and to satisfy everyone. So, there has to be more than enough for everyone, simply to constitute a meal that signifies a feast. It's impossible to signify the same qualities by putting 'polite quantities' or meagre rations on the table. So, festival eating requires, if not 'ruinous feasts', at least feasting. There is no other way – otherwise it isn't festival.

What this means, though, is that where there is festival there will always be waste, for with this kind of abundance of food, there is food waste. In part, this is because of the effects of the contemporary politics of food: putting an abundance of food on the table does not mean that the feast will be eaten in its entirety, most especially in those societies where body thinness is valorized and where disciplining appetite is one of the techniques for governing the self.[35] But it is also an effect of the place of leftovers in culinary cultures. One of my strongest childhood memories of Christmases in the late 1960s, prior to becoming vegetarian, is the seemingly never-ending meals based on turkey. These progressed from the pinnacle that was Christmas dinner (the hot version, with the white meat) and then Boxing Day dinner (the cold version with white meat and lashings of pickles), through various shades of dark meat turkey dinners over the ensuing days between Christmas and New Year, to turkey broth (boiled down bones) and scraggy turkey sandwiches. Eating nothing but turkey for days on end was, of course, a culture grounded in thrift – and in its close associate, relative poverty. But it was, if memory serves me correctly, a real culinary grind. There was nothing fun about this practice of eking out festival food. Which explains why in times of plenty, and in conditions of relative affluence and where there is the option to eat other stuff, this leftover stuff mostly gets thrown away. This is what happens in most contemporary households in the UK, with an estimated 4 million festive dinners being wasted.[36]

An even closer connection between feasting and food waste occurs around celebratory home-based parties. If we think about the typical kids' birthday party, or a celebratory gathering such as a big birthday or an anniversary, it again majors on an abundance of food. But this time it's typically an abundance of what we might term 'junk food' – pizzas, crisps, flans, nuts, cakes and biscuits. Mostly it's also served up buffet style, using disposable cutlery and plates. This kind of food and this kind of eating does not generate leftovers to be eaten on multiple days. Instead, it results in piles of half-eaten, left behind stuff; the kind of stuff that is just tossed into black bin liners and gathered up, when the party's over, as waste, and dumped unceremoniously in the household bin.

Alongside the food, festivals typically also involve some form of gift exchange. The two are not necessarily distinct, for families and friendship groups can, and often do, both take it in turns to host these festival meals and bring contributions to the feast in the form of agreed dishes. But what often happens is that another round of gift exchange accompanies the feasting; this is an exchange of physical goods.[37] Since these are gifts, in many societies they typically come wrapped. Often these gifts come multi-wrapped, contained not just in gift bags but also parcelled up in glitzy, colourful, patterned paper, ribbons and bows – or, overwrapped. No longer is it seemingly sufficient just to wrap a gift; the wrapping itself must be decorated still further.

Wrapping matters. It signals that what is inside is a gift, demarcating it from the things that one might purchase ordinarily for oneself or others, and which remain unwrapped. It also works to mask or disguise that gift, rendering it simultaneously unseen and a surprise. This, too, has effects, most notably in children, whose instinctive response is to rip the surrounds off in their excitement to get at what's inside. The result – as anyone who has dealt with children at Christmas or a child's birthday party will know – is an enormous mountain of paper detritus, enough to fill an entire room. Since it's usually torn and ripped, not much of this stuff is salvageable, to be used again. Instead, it has transitioned swiftly from connoting something special to becoming stuff that itself is of rubbish value. Tidying all this up is inevitably part and parcel of restoring order to a room post the partying. It shows, once again, how discard relates to disorder. It also results in a mountain of stuff being deposited in household bins. So, once again, we see how the bin is the seen to be appropriate location for managing this type of discard, and the means to restoring order to the home.

The festivals that I've looked at so far are those that are celebrated in our homes, involving family and friendship groups. There is, though, another sort of festival worthy of mention that figures across multiple societies. This involves travelling to another site or place and gathering amidst a crowd of strangers. These types of festivals may not involve enormous amounts of food preparation and feasting, but they certainly do result in the accumulation of an enormous amount of waste. They, therefore, show the strong connection between any form of human festival activity and waste generation.

A classic case in point would be an outdoor music festival, something like Glastonbury, which is held most years at Worthy Farm in the south-west of England and attracts over 200,000 people to hear headline acts over three days. Events like Glastonbury give rise to makeshift small cities of tent dwellers – their nearest parallel in living conditions is a disaster relief centre or refugee camp. But these festivals are commercial activities. At Glastonbury 2019, there were 500 food stalls and 900 shops catering for attendees, not to mention 5,000 toilets. Notwithstanding the strong commitment on the part of the event organizers to minimize waste and to promote recycling and renewable forms of energy on the site, the discard generated by event attendees continues to be immense. In 2019, 2,000 tonnes was left behind, including plastic bottles, cans, abandoned tents and camping chairs, along with air mattresses and wellingtons, laughing gas canisters, and food and drink waste.[38] The clean-up after an event on this scale, therefore, is of epic proportions. It also involves labour on a scale rarely seen in relation to waste management in the developed world, for alongside the makeshift city is a makeshift recycling infrastructure, assembled to recover as much as possible from the detritus that has been left behind. So, as the festival goers disperse, some 1,300 volunteers begin to 'pick' the fields, segregating the discard that strews the site into piles of similar materials. Those materials that are potentially recoverable are then transported by tractor to a makeshift materials recovery facility, put together in farm sheds, where some 400 pickers work 16-hour shifts on eight parallel conveyor belts, taking three days to sort the discard into streams of materials recoverable for recycling: paper, aluminium cans, textiles and plastic.[39]

Such accumulations of discard illustrate the connections between mobility and waste. In that regard, they go hand in hand with the 'on-the-go' food and drink example. They also reveal the connection between periods of festivity and acts of discarding. Events like Glastonbury, like all festivals, are transient periods, when normal life for attendees is suspended. As normal life resumes – indeed, for normal life to be resumed – the imperative for attendees is for much of the stuff that is associated with festive life to be abandoned, or left behind. Abandonment signals that these things are no longer needed in the world that is being returned to. But in that very act, these abandoned things become caught in the category of discard. This takes us back to the relationship between discard and displacement. For a temporary festival site such as Glastonbury to revert to its primary use (in this case a farm), it must be divested of all this abandoned detritus. For the social and spatial order of normal life to resume in this place requires not only that the discard of festive life be discarded, but that that discard in turn be ordered – by removing it. This is no small order. The clean-up not only takes some six weeks to execute, it also costs – in 2017, some £785,000.[40]

Discard and the temporality of human lives

Just as discard and discarding mark the temporal ordering of societies, so they punctuate the temporality of human life itself. There are multiple examples of this.[41]

Perhaps the most obvious is the clearing of effects that accompanies every human death. This process is defined and shaped by categories of things, and the sorting of things into these categories. There are those things that are to be kept, as mementos of a life and of personal attachments; the things that are given to others and, if the deceased is a parent, the process of dividing up certain things between siblings. Then there are the things that the deceased valued but that those who remain do not and working out what to do with these. And then there is everything else, or the stuff of ordinary life that is left behind, the remainder of a person having lived in a place. These are things like the clothes, the cooking utensils and the cutlery, the glassware, plates, the furniture, fitments and furnishings; and then the house itself. Dealing with some of this discard, most especially furniture, takes a lot of time, especially if one seeks to direct it through conduits and channels that will result in its reuse and reappropriation. Where time is in short supply, therefore, or where the motivation to work through all this stuff is lacking, or where there are no close kin relatives living close to hand, then the tendency is to use 'house clearance' agents, who, for a fee, will remove everything and use their extensive contacts to supply the secondhand market. In such a way, discard can be repurposed as 'pre-loved' or 'new to you', sold on eBay, or, if it is more valuable, sold on to a specialist dealer. Either that or the temptation is to hire a skip for a weekend – a course of action that results in a large amount of stuff being jettisoned, and then becoming waste.

If death is the most obvious point at which discard manifests itself as an effect of life, there are other close parallels. Moving house is often similar in the processes enacted in things. Anyone who has moved house after spending several years living at the same address will find themselves going through an identical process. There will be the stuff that is definitely going to the new house, but equally there's a whole lot of stuff that might not be suited to the new location or that won't fit in. So, there is the challenge of what to do with that. And then, if one is a parent whose children are adult and have left the parental home, there is often all that stuff that's in the loft – an archaeological mound of children's material culture, typically 20-plus years old and of no value to anyone now. So, with time running out, and a hard deadline of a moving date, even the most committed of recyclers and thrifty consumers can find themselves having to resort to what are described as 'binning and tipping' practices simply to manage their no longer wanted stuff.

The same process gets enacted in university towns and cities across the UK each summer as students graduate and move out of private rented accommodation, except this is conducted to an even tighter schedule than moving house. In itself that means more stuff is discarded, for the pressure of the move-out date is felt in more stuff falling into the category of abandoned stuff. But more than this, this form of moving out is also a key passage point in life – a point where a student life is left behind, along with all those student things. The cheap desk, the student mattress, no longer wanted clothes. While some may endeavour to sell some of these no longer wanted things (particularly textbooks), others merely jettison them, often leaving them in piles in yards to be collected (usually days later) by local authority student clean-up teams.

Discard and discarding, then, figure at key transition points in life (and death). They also characterize those points when living and working arrangements change. Think, for example, of when cohabiting living arrangements break down. In these situations there is, again, the same division of stuff into key categories, inflected in this instance through the 'whose is what?' process. But often in these circumstances there is also a jettisoning of stuff, which may also involve sabotaging certain things, most especially if the break-up has been acrimonious. At work here is exactly that expulsion imperative that we saw previously. Jettisoning here marks the need for a subject to reassert their identity, separate from things that are too-potent reminders of previous attachments. It also clears the ground for potential renewal, in the form of new attachments. Sabotage shows an even stronger imperative, one of displaced violence, in which the thing is the substitute for the person. In both cases, it matters that the stuff becomes waste, for trashed things, and trashing things, do important work for the restoration of subjectivities.

A rather different, mostly less acrimonious experience is that of clearing out an office or place of work when one leaves a certain sort of professional job or retires. The process, however, is much the same and one enacted through exactly the same categories in things. In the last decade, I've done both these things, and the process has been pretty much identical both times. There is the same separation of that which is to be kept and can be kept (and that still has value) from that which is to be abandoned and left behind. Mostly, for academics, this category comprises large numbers of no longer wanted and/or needed books as well as other types of paper – printouts, photocopies, and so on. Typically, such stuff is made available to those who remain in the workplace on a 'free for all' basis. Then there is the mountain of other stuff that cannot be simply abandoned but rather is discard that must be directed in particular trajectories. There are the piles and piles of paper, filing cabinets full of it, which summarize years of meetings, the policy process and policy decisions taken – and taken again. These records are rendered valueless on leaving that workplace; they are of value only as recyclable paper. But there is also the discard that must become confidential waste, for secure shredding, for example, piles of papers that relate to confidential assessments and past audit exercises and materials from research projects that could be connected to particular individuals.

Finally, discard and discarding punctuate the more everyday rhythms of how humans live with stuff. This is because everyday life is full of ordering activities. Consequently, everyday practices of living, or the sheer ongoingness of life, all result in things being routinely discarded. These practices could be as big as redoing a kitchen or bathroom or as small as changing the winter clothes in a wardrobe over to the summer clothes, or even caring for small children's clothing through the never-ending practices of washing and ironing. Invariably, moving clothes around by doing the laundry or changing over wardrobes results in their evaluation: does this garment still fit? How much more wear would a child get out of it? Can it be saved for a younger sibling, or would it be better being handed on to a friend with a slightly younger child? Is this garment looking 'tired'? Can I get away with wearing this again this summer? Oh no, not that again – I'm fed up with that! Questions like these are ones of valuation and they sit behind clothing discard. But this kind of clothing discard is rarely binned. In other words, it seldom directly becomes waste. Instead, it passes through conduits as varied as eBay, a charity bag drop collection, charity shops, swap shops and family/friendship networks, through which and around which small children's clothes circulate intensely. This is something that I explore in more depth in Chapter 4. Redoing a kitchen or bathroom, however – something that occurs frequently when people move home or when people seek to give their existing home a new lease of life – generates large quantities of stuff that is directed mostly towards the waste stream. Old toilets, sinks, baths and basins, tiling, carpet or floor coverings, worktops and unit doors (if not the frames) are all typically dumped at household waste and recycling centres, with very little of this stuff being salvaged.[42]

Where there is life, there is discard – and its implications

The foregoing makes clear that it's not just that discard and discarding are central to the human psyche and to the temporal ordering of human societies; it's even more elemental than that, for where there is human life, there is discard. As I have shown in this chapter, the act of discarding is fundamental to how humans live in the world. Discarding figures as a constant motif in our everyday lives. It punctuates the major stages of the life course; it marks key passage points and rupture points – when our social relations break down and living arrangements change; but it also marks key points of renewal and return. We discard in order to make room for the new and to return to everyday life after periods of festivity. Discarding, then, is thoroughly entangled in how we make sense of ourselves as social beings in social worlds. This is because, unique among species, humans enact social relationships through things. Our cultures are intensely material.

All this is essential to recognize. It also has considerable implications. To spell it out, it means that it is impossible for there to be a world that involves humans that is without discard and discarding. It is unrealistic, therefore, to approach the question of waste by trying to eliminate or minimize discard. This just won't work. Instead, a more promising line of thought to pursue is to focus on the distinction between the categories of discard and waste. That distinction is always a contingent matter. So, there are conditions in which discard remains as discard and is therefore open to being reappropriated, and then there are those conditions where it rapidly becomes waste. Frequently, this is less about the stuff of discard but rather more a matter of the intersection of time with the quantity of stuff in the category of discard.

I want to close this chapter by highlighting that looking at discard and acts of discarding tells us that discarding is a 'lumpy' practice. It comes in peaks and troughs. While discarding is a constant accompaniment to the practices of everyday life, and everyday life results in routine volumes of discard across much of a year, there are also periods in all societies, notably relating to festivals, where large numbers of households produce large volumes of discard simultaneously. This synchronicity presents problems for the waste management industry, with Christmas, for example, being identified as responsible for a 30 per cent surge in the normal volume of household waste across much of the developed world.[43] This poses the industry major challenges, most especially for collection and processing infrastructure, where operational capacities are greatly exceeded, often with knock-on consequences. Overflowing bins and piles of 'side waste' are but some of the problems that face collection crews, while sorting technologies have to be run harder for longer to cope with the surge in volume, with inevitable effects as machinery parts break down and wear out. There are also economic challenges. A 30 per cent surge at Christmas presents the waste industry with the equivalent of the harvest 'glut': a surfeit of material. Further, with limited capacity to store its outputs and the need to sell into the markets as they come on stream, the effect inevitably is an excess of supply over demand, and low prices. For the waste management industry, then, harvesting Christmas is more 'community service' than commercially lucrative. That is perhaps worth bearing in mind as the three Rs (reduce, reuse and recycle) are increasingly being highlighted to consumers as how to do Christmas. As this chapter has shown, however, to attempt to redraw Christmas through the three Rs is to fundamentally misunderstand the social work of Christmas. Life, and most especially celebratory forms of life, goes hand in hand with practices that result in waste making. If we are to understand waste, we need to understand this.

3

Consumption, Consumer Practices and Consumer Discard

Or, how consumer discard relates to economies

In this chapter, the focus shifts from recognizing that discard is foundational to being human, and thus unavoidable, to seeing how consumer discard, and acts of discarding, are baked in to many economies.

In this regard, I want to go beyond the largely theoretical accounts in the social sciences that see discard as an inevitable aspect of economic organization, particularly under capitalism. To be clear, I am not dismissing the power of these arguments. There is much that is of merit in them. Not least is that they establish key general principles that help explain why waste is an endemic part of capitalist economic activity. I begin the chapter, therefore, by outlining these approaches and how they see waste, before going on to show why these accounts are less helpful to understanding consumer discard.

Marxist and neo-Marxist readings of economy recognize that all primary and manufacturing processes involve transformations of the material world brought about by (human) labour, and that they also entail material transformations. But, just like the firms they analyze, mostly these accounts focus on the end point of the production process: the commodity. This is because here, labour (and the commodity that is the result of production) is seen to be the means to value creation for capitalist firms. Yet, in highlighting the overarching significance of commodities, these accounts also flag the relative insignificance to many capitalist firms of the inevitable material leftovers and residues of production processes – or the stuff that, while integral to production processes, never realizes the commodity form. They also flag that residues have mostly been seen as waste, that is, as stuff of zero value. This helps to explain why the history of how firms have dealt with residues is characterized by 'dump and disperse', and never mind the consequences. When stuff is seen to be of zero value, it is but a short step to that stuff being discarded, and discarded as waste. Examples of the residues of economic activity that over time have been dumped in this way include the tailings that come from mining operations and ore preparation; the husks, shells, skins, stones and pips that are stripped from agricultural crops prior to their becoming an agricultural commodity; and an array of remainders from the various chemical and industrial processes that produce basic feedstocks for core areas of manufacturing.

Relatedly, there are those accounts that highlight the capitalist imperative for what is often termed, after Schumpeter, 'creative destruction' and its connection to the renewal of conditions conducive to capital accumulation. The term 'creative destruction' applies not only to firms but also to both consumer and capital goods, including industrial plant and equipment. For value to continue to be realized requires that new goods replace old, be that a cheap consumer good, such as a £2 t-shirt, or a capital good, such as an oil tanker. Replacement takes place over a time spectrum; at one end would be the buy-today-throwaway-tomorrow t-shirt, but the oil tanker would be at the other, with a working life in the range of 12–25 years. In both cases, the surest way for replacement to happen is for the old to be scrapped, or for its form (as an object or entity) to be destroyed. The most extreme versions of these arguments about creative destruction see war as underpinned by economic conditions, if not economically determined.[1] Rubble and ruin are seen to provide the opportunity to recharge stagnant economies.

A less extreme form of the same argument appears in what passes under the phrase 'economic restructuring'. Enduring geographies, in the form of long-established spatial arrangements and geographical configurations of economic activity, are eventually challenged, be that by 'capital flight' (the decisions of transnational firms to relocate to more profitable locations) or through the effects of regulation, notably state policies in relation to regional policy say, or a return to free market economics.[2] Those who live in regions that have experienced the sharp end of industrial restructuring, such as the old manufacturing regions of the UK or the US 'rustbelt', will have seen this for real in the early 1980s, with the closure of mines, factories and assembly or production plants and the attendant widespread loss of employment. This was accompanied by the mothballing or demolition of capital plant, and the frequent abandonment of these places as sites of capital accumulation, with devastating consequences for the livelihoods of people living in them.[3] A more recent and ongoing example of capital restructuring is the decimation of 'bricks and mortar' retail concentrations consequent upon the rise of online retailing. Whether town and city centre retail stores will be demolished, or turned to an alternative use, through conversion to residential use say, is currently an open question.

Then, in work of a rather different theoretical stripe, there is the body of research on innovation and its connection to the business cycle. Innovation rests on continued development, be that of products, processes or organization. In so doing, it produces, as an inevitable effect, obsolescence. Obsolescence applies most obviously to goods and products, and some of it is built in – think of product 'upgrades' and new releases at one end of the spectrum, through to the advent of new technologies at the other. A good example of how technologies become obsolete would be landline phones and the now widespread redundancy of the infrastructure of the public payphone box consequent upon the ubiquity of the mobile phone. Obsolescence also applies to production processes. Continuous improvement endlessly seeks further efficiencies from production lines, most notably currently through the increasingly widespread application and roll-out of robotics. In many cases, this is reconfiguring not only a plethora of jobs but also the built form of working environments, as organizational arrangements designed to maximize efficiencies in relation to human labour are reworked in relation to post-human labour.[4]

Across the canon of work in the social sciences, there is much theoretical work that alerts us to the economic necessity of discarding under capitalism. The same was true under state socialism. Wherever there is economic activity, then, there will be abandoned stuff as an effect of production processes, and in capitalist economies additionally as an effect of the cycle of innovation that drives the product cycle and as an effect of the declining profitability of production in certain places over time.

The trouble with these accounts, however, is that they are just that – general accounts. They reduce everything to, and explain everything in terms of, general principles. What they miss, therefore, is the contingencies of the economies in which people live and work. These are particularly important when it comes to understanding consumer discard. In many economies since World War II, the bare facts are that consumption has played an increasingly significant role. This can be seen in direct employment statistics and in the contribution to economic output of sectors that depend on consumer spending – sectors such as retail, food and drink, hospitality, personal services and tourism. Consumption also underpins the private housing market. Not only is this a key engine in driving much consumer purchasing, it is also entangled in a trend now identified across multiple advanced economies, namely, the switch to asset-based accumulation, particularly through property ownership, and its attendant connections to widening inequality.[5] No better illustrations of consumption's economic significance exist than the G7 governments' responses to the global financial crisis (2007–9) and to the COVID-19 pandemic. Both showed that keeping particular economies afloat depended on creating the conditions that allowed for consumers to keep being able to service their mortgage debt and to keep on spending on consumer goods and services.

Consumption's significance to economies has some quite profound effects on consumer discard. At the level of individual economies, there is a clear trend. As levels of prosperity in countries increase in response to growth, and as public policy levers are pulled that encourage consumers to increase spending on consumer goods, so the volume of consumer discard also increases. With that relationship goes a change in the type of consumer discard that is generated and its material composition. This is what lies behind the changes in household bin contents that are observed across the 20th century in the US and Europe.[6] Discard ceases to consist primarily of ash and organic scraps and instead starts to encompass consumer goods, and then more and more consumer goods. One interpretation of that transformation has been to see such discard as emblematic of consumers' materialism and the endless desire for the new; another has been to label it as a manifestation of a 'throwaway society'.[7] Both interpretations effectively lay the blame for consumer discard with consumers. They see it as a problem that is either psychological or social. In this chapter, I show that such accounts are pretty wide of the mark. Instead, I will argue that consumer discard is better understood as the residue of economies that rely in large part on consumer spending. So, I see consumer discard largely as an effect of capitalist economic organization, although, as will become clear in later sections, discard is also thoroughly bound up in social life, most notably the ordinary practices of consumption that orchestrate and order everyday life.

To begin to develop this argument, I turn first to expand on consumption's significance to economies and then to how consumer discard is an inevitable part of such economies.

The emergence of consumer-heavy economies

The importance of consumption in and to the economies of the global North can be traced back to the Great Depression of the late 1920s and 1930s in the US and Europe. The challenge of those times was the inability of the then dominant free market economics to provide public policy solutions to desperate economic circumstances – mass unemployment and business collapse. Enter John Maynard Keynes. In simple terms, Keynes argued that demand mattered more than supply for economic growth, specifically that the key economic driver is aggregate demand, comprising the sum of spending by households, businesses and governments. Prolonged unemployment, he argued, was the result of inadequate demand. In recessions, demand fell first, as a response to contracting consumer spending. This, in turn, led to less business investment; job losses and – if the spiral persisted, uncorrected – mass unemployment, of the level seen in the Great Depression; and business collapse. Keynes's suggested public policy response, first enacted in the US through Theodore Roosevelt's New Deal, was to boost government expenditure, chiefly via large infrastructure projects that acted to kick-start employment and, through that, to stimulate consumer demand. The world's first genuine consumer economy and mass consumer society, therefore, was born out of the depression of the 1930s and the public policy response in the US.

The subsequent US recovery and economic boom meant that Keynesian economics came to define economic policy across what were termed the 'advanced economies' in the post-World War II era. As a consequence, consumption came to be seen as central to these economies, with policy levers being used by governments to stimulate consumer demand. Increasingly, that came to mean persuading consumers to shift from time-honoured savings-based cultures of purchasing, to be willing to incur debt to purchase goods. This occurred most notably through the expansion of mortgage lending to finance house purchases but also through hire purchase agreements for many consumer goods. In such a way, the consumer revolution that took the US by storm in the 1950s spread to Western Europe in the 1960s, with these decades seeing the widespread adoption of a suite of 'consumer durable' goods. The effect was not just societal, for stimulating consumer demand also worked to expand employment in the manufacture of consumer goods. These connections are usefully explored by another Keynesian, JK Galbraith, in his book *The Affluent Society* (1958).

Galbraith's starting point is with the two underlying pillars in economic theory for understanding consumer behaviour. These are the theory of 'consumer demand' and the theory of 'marginal utility'. Both seek to articulate the relationship between human needs, wants and desires. As basic living standards rise in an economy, and as per capita real income increases, so the importance of needs (chiefly food and shelter) recedes, to be supplanted by wants and desires (in the parlance of economic theory, utility). The theory of consumer demand states that the urgency of wants does not diminish as more are satisfied. Rather, wants beget more desires, so desires can never be satisfied. Insatiability is compounded by emulation – more people desire the same goods, thereby expanding the market for those goods. The theory of marginal utility provides a degree of elaboration. It states that desire is a function of the quantity of goods available to an individual to satisfy that desire, so the larger the stocks of goods available, the less the satisfaction. Galbraith's addition is to argue that wants are dependent upon the process by which they are satisfied, or 'the dependence effect' (pp 135–6). He maintains that it is the production of goods (and their advertising) that works to induce wants, and that wants in turn work to drive more production of those goods, and thus more wants. So, what happens in an industrial economy where consumer demand is stimulated is that the balance of employment starts to shift from a reliance on the extractive industries and manufacturing output for industry, to the production of consumer goods, to ensure the continued satisfaction of consumer demand. Employment and economic output start to become consumer heavy, dependent on continued consumption.

This is what happened across the advanced economies in the period through to the 1970s. The pattern of employment in these economies turned increasingly away from the heavy manufacturing industries (coal, steel, ship building, chemicals) to the production of cars, an increasing array of consumer durables (fridges, washing machines, vacuum cleaners, and so on), furniture and furnishings, and processed food, with large numbers of jobs dependent on each of these sectors. As early as 1958, for example, one in six jobs in the US were in automobile manufacture, and as late as 1972, over 500,000 were directly employed in the UK car industry.[8]

Subsequently, the stagflation of the 1970s brought about the loss of sway of Keynesian economics in public policy and a return to free market economics in the US and UK under the Reagan and Thatcher administrations. In turn, that fuelled the rise of globalization and the collapse of an economic geography that saw domestic production meeting the demands of domestic consumers. A trend that began with the car industry – in which, in the 1980s, better-quality Japanese cars captured an increasing share of the markets in the US and UK – extended to encompass other higher value consumer goods, notably TVs and consumer electronics. Then, as capital chased ever cheaper sources of labour, the production of lower value consumer goods, especially footwear and apparel, got outsourced to the factories of South China and East Asia. In 2009, with the relocation of their factories to Poland and the Czech Republic, Hotpoint and Hoover – the final remaining major washing machine manufacturers in the UK – ceased UK-based production, resulting in thousands of job losses in Wales, while only 169,000 are currently directly employed in car manufacture/assembly in the UK.[9]

In today's world, the dependence effect identified by Galbraith, therefore, is no longer national in scale but global, with the wants and desires of the consumers of the global North met through largely Asian labour. That said, the economies of the global North still retain a reliance on consumer demand and consumption-related employment. This version of the dependence effect is less about the manufacture of consumer goods and, rather, more about purchasing consumer services, particularly in the form of tourism, hospitality, entertainment, leisure and experiences. In the UK alone, for example, in 2019, some 2.5 million were employed in accommodation and food services, while hospitality accounted for 21 per cent of tertiary sector employment.[10] More broadly, there has been no better indication of the consequences of the dependence of European economies on consumer services than in the response to the COVID-19 pandemic, where desperate efforts to keep the tourist industry afloat saw political deals struck to allow tourists to flock to Spain and Greece in the summer of 2020, with knock-on consequences for virus circulation. Similar stimulus packages were pushed within economies. In England, for example, the exit from the first lockdown in summer 2020 was marked by a government-financed 'Eat Out to Help Out' campaign, which encouraged consumers back to the hospitality industry in their droves through heavily discounted meals. Over a decade previously, as the G7 response to the global financial crisis took shape, that same recognition of the imperative to keep consumers spending in order to stop economies tanking again took centre stage, with enormous stimulus packages designed to ensure that consumers did not default on their mortgages.

Since the 1950s, then, consumption and consumer demand have become baked in to the economies of the global North. The Keynesian legacy is that the success of these economies in terms of their growth is reliant upon continued consumer demand. More than this, much of their employment is dependent upon people having the money to continue to buy not only consumer goods, but also consumer services. It is not too much of an exaggeration to say that the economic well-being of many households in countries like the UK is entirely dependent upon consumer spending, and thus that economies like this have become, if not consumer economies, certainly economies that depend on creating the conditions for sustained, continued consumption.

Beyond the global North, in China, Deng Xiaoping's reforms, the transition to market capitalism and the promotion of consumption through the 'get rich first' policy have meant that a parallel transformation has occurred there since the 1990s.[11] A trajectory that took the best part of 50 years to solidify in countries such as the US and UK has occurred with typical compression in China. In the 2000s, home ownership expanded rapidly, fuelling a boom in demand for consumer durables and furniture and furnishings. At the same time, demand for other key consumer goods – cars, mobile phones and clothing – increased inordinately. To this has been added heightened demand for key consumer services, notably tourism, entertainment and hospitality. What has been labelled as 'materialism with Chinese characteristics' has seen a degree of tempering recently. There has been a growth in the secondhand market, particularly online, and the emergence of leasing models in relation to the market for luxury goods.[12] To that can be added appeals to the virtues of frugality, especially in relation to eating out. But this is largely tinkering around the edges. Domestic consumer demand is now a primary driver for the Chinese economy. China is no longer simply the 'factory for the world', although it continues to be a major supplier, accounting, for example, for 47 per cent of global textile output in 2017.[13] Rather, as the Chinese economy moves up the value chain, the motto is more 'made in China, sold in China', with the clear intent being that China can satisfy an increasing share of consumer demand through domestic production. That aspiration remains to be demonstrated, given that Chinese consumers are significant purchasers of luxury global brands. Nonetheless, China is widely seen by market analysts as the world's fastest growing consumer market. In 2019, McKinsey stated that China now accounts for over 18 per cent of all final goods consumed in the world. A further indication of the spending power of this market comes from 'Singles Day' 2018 – a one-day e-commerce bonanza that saw sales of $125 billion, exceeding the combined totals of Black Friday and Cyber Monday in the US.[14]

Beyond China, consumer demand has become a key plank of economic policy across much of the world, not only for the major emerging markets such as India, Brazil and South Africa but also for middle-income countries such as Mexico and Bangladesh, whose economies previously had functioned largely as assembly factories for goods designed to be consumed in the global North. Mostly, the size of this emerging new middle class is estimated at around 30 per cent of populations. Recent estimates of global inequality flag their significance to the balance of global consumer spending power.[15] They highlight that the primary beneficiaries of 'high globalization' (1988–2008) in terms of rising real incomes were very much the emerging Asian economies (China, India, Thailand, Vietnam, Indonesia). They also show the stagnating spending power of the lower middle classes in the 'Old Rich World' – the US and Europe – as well as the gaps that are emerging in these countries at the lower end of their income distributions, between the 'haves' and the 'have nots'. While the economic downturn post the 2008 financial crisis dented growth in the emerging economies, and thus dampened down consumer demand, their recovery has been stronger than in the Old World, especially in comparison to countries that turned to fiscal austerity, notably the EU bloc. Consumer confidence has returned in these countries and the demand for goods and services continues apace.

All this has major implications for thinking about the connections between consumption and discard. A simple economic truth is that any economy that is consumption heavy is intrinsically an economy in which the ongoing turnover and replacement of consumer goods comes to matter for economic growth. The significance of consumer demand to the economies of a rapidly growing array of countries, then, means not only burgeoning consumer markets in many parts of the world, but also an expansion of consumer discard as more and more countries face the challenges of dealing with the economic inevitability of product obsolescence.

Product obsolescence and consumer discard

As a term, 'product obsolescence' encompasses two dimensions: obsolescence that is the result of a loss or impairment of function, and obsolescence that connects to the perceived desirability of a good. Whereas the first is a matter of the capacities of something to continue to do what it was designed to do, the second is frequently, although not exclusively, linked to fashion – in the sense of the new or latest things – and technological innovation.

Technological innovation and advances mean that the capacities of things continue to be developed. The consequence is that things that continue to function but that have been supplanted by more recent models, and most especially new technologies, become, to a greater or lesser extent, less desirable. Cross-cutting this is how obsolescence connects to the key drivers in capitalist economies. These connections were first articulated by Bernard London in a pamphlet that appeared in 1932, tellingly entitled *Ending the Depression Through Planned Obsolescence*. The same connections then featured heavily in Vance Packard's 1960 key book, *The Waste Makers*.[16] In conditions when firms' profitability depends on consumers spending, it is no longer sufficient simply to expand the geographical reach of the market and the degree of market penetration. Consumers need not only to continue to buy goods, but a good number of them also need to replace goods relatively often. In these circumstances we start to see firms resorting not just to fashion but also to strategies of manufactured and planned obsolescence to ensure continuity in sales. A set of examples, each of a particular category of consumer good, illustrate these general processes.

First, let's take a look at the category of household 'consumer durables', or goods that, by definition, are meant to last longer than others. The focus is on two 'white goods': washing machines and fridges. In both cases, I take the long historical view, going back over three generations, rooting this in my family's history in these consumer durables. I do this not only to establish how obsolescence has evolved over the period from the Great Depression, but also to tell a social history story that illustrates the gendered history of domestic technology.[17] The latter foregrounds a later concern of the chapter, which is with how consumer goods relate to consumption as this is practiced in the home.

If I think back to my childhood, my mother's kitchen in the 1960s included what was seen at the time as the latest 'must-have' clothes washing device: a twin tub. Why was this device so desirable? Quite simply, because it was the first real labour-saving laundry device made available to 'the housewife'. By contrast, the washing at both my grandmothers' houses, even then, was done by hand. In their houses, much like it had been for generations of working class women before them, Monday was wash day. Much like it had been for their mothers, it involved standing at the kitchen (or scullery) sink most of the day, scrubbing garments with hard soap, rinsing them and then running them all, individually, multiple times through a hand-turned mangle, before hanging them out to dry on a washing line. This was hard, physical work – particularly with bedding – and it took its toll on hands, which routinely suffered from chilblains and dried skin conditions. The twin tub, by contrast, took the physicality out of washing by automating the tasks. It heated the water, it washed the clothes and it spun them dry enough to hang out on a washing line. But, as I recollect, the device still needed a lot of supervision, for it had to be attached to the sink taps in order to fill with water, its integrated spinner was turned on and off by hand, and emptying the tub of heated water involved making sure that the drain hose that was hooked over the sink did not come adrift and flood the kitchen floor. The last was the job of child labour! The twin tub, in other words, may have been a labour-saving device, but it still required an operative – the housewife. Come the 1970s in the UK and the housewife operative was getting to be in shorter supply, as women's participation in the labour market increased. Cue the advent of the plumbed-in, front-loading, automatic washing machine – a revolutionary device that, once loaded, could be left to do the washing unsupervised. This device rapidly achieved near universal adoption in UK households, replacing twin tubs, hands and mangles as the laundry technology of choice. The latter devices were widely declared to be of rubbish value and hence got discarded. Significantly, however, this is the last such occasion in the development of washing machine technologies. Albeit that the capacities of washing machines have increased over recent decades, including the development of integrated washer-dryers, low-temperature 'environmentally friendly' washes, delicates care and multiple spin options, none have persuaded consumers that they need to replace their machines in droves. Desirability has not overcome function sufficiently for replacement to occur prior to the breakdown of the machine.

In these circumstances, if manufacturers are to continue to sell products, then function has to begin to fail. This is what is known as manufactured obsolescence and it is what has happened with many brands of washing machines sold in the European market, at least judging by the evidence, which points to as many as 40 per cent of households having to replace a washing machine within six years.[18] This connects to the economics of design and manufacture. While there are, undoubtedly, brands of washing machine that are manufactured to last, these command a high price point and are beyond the financial means of many households. By contrast, in manufacturing a good to be sold at a relatively affordable price point, what matters is cheap parts and efficient manufacturing and assembly processes. Cheap parts and efficient manufacture/assembly result in appliance failure. They also mean that, when washing machines break down, they are replaced rather than repaired. The economics of repair here are such that repair is an uneconomic option for the consumer, for to repair them would often require that the washing machine be pretty much manufactured again. So, the rational economic choice for consumers is to replace the good. What might look like a situation in which lots of people are choosing to get rid of washing machines, then, is more an effect of sound household financial decision-making.

Fridges are a rather different case. Again, if we take the long view, we can see how the technology has evolved. Unlike the case with washing machines, both my grandmothers had fridges in their kitchens in the 1960s. The presence of this device was what marked their kitchens as different from those of their mothers, who had had larders – a small storeroom off the kitchen in which they kept stocks of dry goods like flour and foods they preserved annually such as jams, plums and chutneys, pickled onions and beetroots. By contrast, fresh food – chiefly meat, milk, eggs and butter, but also cheese and many vegetables – could not be stored at room temperature for longer than one or two days. This meant that in the early decades of the 20th century, daily food shopping was essential in order to feed the household. Fridges started to become affordable for UK households in the 1950s.[19] The immediate benefit of the fridge to the housewife was that it stored food safely for longer, thus reducing the necessity for daily food shopping. Fridges, along with vacuum cleaners, are generally regarded as the first of the labour-saving consumer durable goods, but their food storage properties mean that they are also bound up in the development of food manufacture and processing. In the post-World War II period, with the advent of food manufacture and processing, came the introduction of frozen foods – not just frozen cuts of meat and fish and packs of vegetables but ready-made goods sold frozen and ready to heat, such as fish fingers, pizzas and burgers, and then ready-made meals.

Over time, and particularly as women's participation in the labour market grew, as supermarkets captured an increasing market share of food sales and as shopping patterns changed from multiple times a week to weekly, the size and shape of the fridge changed to accommodate the need to store more food. The standard model was no longer a small rectangular box with a very small ice box – sufficient to hold one family-sized pack of frozen peas. It became a much taller box – the integrated fridge-freezer, with varying sizes of fridge and freezer component to suit household circumstances. This device became the default purchase for working families, for whom it offered convenience. It allowed them to shop weekly, but also, in doing that, it freed up time that would otherwise have had to be spent food shopping for doing other things. So, the fridge-freezer is not just about the time-saving sense of convenience, it's also what's known as a scheduling device (Chapter 1); it allows for the coordination of multiple people, in this case doing things other than food shopping. Accommodating fridge-freezers in UK home kitchens, however, was most definitely a problem. Many kitchens got redesigned to incorporate these new devices, with the result that much of the existing kitchen infrastructure, especially wall cupboards and work surfaces, ended up being discarded and replaced. This connects to a phenomenon known as 'the Diderot Effect'.[20] So, we see here how accommodating a new, seen to be essential domestic technology can result in large amounts of different, but related, consumer discard.

For the last 20–30 years, not much has changed with domestic fridge-freezing technologies. There have been periodic efforts by manufacturers to stimulate product replacement through appeals to fashion such as stainless steel, shiny-surface fridges or coloured fridges. More recently, there has been the advent in the European market of the 'American' fridge – a vast double-doored device, with a water cooler and ample space for keeping beers cold. This again requires a major kitchen redesign to accommodate, but its size also signals the 'overconsumption' of food, and while cold beer storage capacity might sell well in Arizona, it's perhaps got less of a market edge in parts of Europe. So far, none of these innovations has achieved the levels of market penetration associated with the advent of the fridge-freezer. And, with less moving parts, the durability of these products is relatively high: a turnover time of ten-plus years is not unusual, even with mid/low-price brands, while the top-end brands are likely to last 20 plus years.[21] So, fridge-freezer technology poses something of a problem for at least some manufacturers – it's so good, it lasts! This, however, may be about to change.

The arrival of the 'internet of things' in homes has led to the development of the smart fridge-freezer. As with the fridge-freezer, the key innovations that accompany these devices enable enhanced time-space coordination – something that can be anticipated to appeal strongly to those trying to choreograph busy family households. Its capacities to connect with a mobile phone allow for the remote checking of household food stocks while shopping; its capacity to date-scan the food being stored and to connect to potential recipes is a means, potentially, to order household food consumption and reduce food waste; while its capacity to display and integrate household members' calendars provides a digital counterpart to the fridge door's role as the family's analogue-mode coordinator. That said, smart technologies will only remain smart for as long as their operating systems are supported, and while Apple products are guaranteed to receive that support for five years, Samsung's and other Android devices only get full updates for two years. So, a smart fridge could very quickly become not only a 'dumb' fridge, but a major security risk, with the implication that the replacement cycle for domestic fridge-freezer technology might just have got a lot shorter. Smart fridge-freezer technologies are an illustration of what is meant by planned obsolescence. While they might have a positive effect on manufacturers' bottom lines, their effect is likely to be considerable in terms of consumer discard.

If the manufacture of consumer durables is just beginning to illustrate the principles of planned obsolescence, another category of consumer goods provides examples that exemplify them: home entertainment goods. The classic good here is the TV.[22] Again, my family history provides a window to some early universal patterns. So I am told, we did not have a TV until the early 1960s. This was a small black and white set, which we got secondhand from friends of the family who were upgrading. That set didn't last long before it broke down, irreparably apparently. But, having grown accustomed to TV watching in the evening once we kids were in bed, my parents decided they needed to replace it. Unable to afford a TV purchase, like many households at the time they resorted to renting a TV from a company that became a household name among our social class: Radio Rentals. Visits from the TV repair man (invariably wearing a white lab coat, as the badge of technology) punctuated my childhood, as the set varyingly failed to tune to the few available channels, failed to keep a steady picture or, more spectacularly, blew up, as it did on one occasion, emitting flames from the back, and was promptly doused with buckets of water by my mother (after switching it off first!). In 1969, however, the family splashed out and bought a colour TV. The precipitate to this purchase was the Apollo 11 moon shot. Again, this was at my mother's insistence – this event was history in the making, whatever happened, she reasoned; we had to engage with this in full colour. So, along with the rest of the world with access to a TV, there we were on the edge of the settee, holding our breaths, looking through the fingers of our hands as we heard and then watched The Eagle land on the moon.

Colour cathode-ray TVs were the stable TV technology for several ensuing decades, with little other than screen size differentiating models. As I recollect, too, the sets were reliable – gone were the days of the need for being on friendly terms with the TV repair man, and replacements were few and far between. Then, in the early 2000s, affordable flat screen technology (plasma and LED) was applied to TVs. At the time, I was conducting a research project on household discard, and the effect of this change was impossible to exaggerate. Household after household participating in the study started jettisoning their cathode-ray tube TV and replacing it with a flat screen TV. This was the equivalent of the fridge being replaced by the fridge-freezer, or the adoption of the front-loading automatic washing machine, and it was further accelerated by the switch from analogue to digital transmission. It is another example of how major technological advancement suddenly renders obsolete a whole category of good, with huge consequences for turning those goods to goods of rubbish value.

The period since the early 2000s has seen TV product development oscillate between the function and fashion dimensions, but increasingly through the prism of the home as a growing site of entertainment. This has been a key trend in home-based consumption in recent years.[23] There have been appeals to an enhanced visual experience, initially through HD and more recently 4K- and 8K-enabled sets. Typically, these roll-outs have been linked to the optics of viewing a key global sporting event, for example, the football World Cup and the Olympics. Then there has been the advent of the smart TV. This more-than-a-TV is an entertainment platform that, through its streaming capacities, allows the TV to become the means to access any form of video content, be that subscription services such as Netflix, gaming or live-streamed, pay-to-view entertainment. Neither of these developments, however, has yet resulted in the kind of wholesale replacement of TV technology that occurred with flat screens. Instead, there has been the gradual adoption by those for whom these viewing experiences and viewing choices matter, accompanied less by the jettisoning of previous models but rather the proliferation of TVs in the home through the trickle-down effect.

If we look inside contemporary UK home interiors, the presence of TVs in rooms has changed markedly since the 1960s. In the 1960s and 1970s, and even into the 1980s and early 1990s, the TV was the centrepiece of the family sitting room, around which everyone gathered to watch. Now TVs are likely found in the kitchen, in family rooms, in children's and parents' bedrooms and, increasingly, at least in larger houses, in rooms dedicated to home entertainment that are the domestic equivalent of the cinema. As the functionality of smart TVs becomes more habituated within households, an open question is to what extent trickle-down will continue to satisfy the increasingly individuated viewing practices of household members, or whether streaming to a mobile or tablet device will function as an appropriate substitute for the big-screen home entertainment experience. The alternative is that home entertainment becomes not just the equivalent of the cinema, but that more rooms in a house are fitted out with smart TVs, in the process turning the home into the equivalent of the multiplex, in which case, large numbers of 'dumb' TV sets would be rendered obsolete and turned to discard.

Even more intense cycles of technological redundancy characterize information and communication technologies. Home computers, laptops and tablets, mobile phones and gaming devices are the archetypal goods in this category. While these goods rarely suffer from hardware failures, they are replaced frequently. This is an effect of changes to operating systems and in device capacities. So, let's take home computers. Replacements here are mostly an effect of big tech's decisions with respect to their operating systems. While some of these are incremental, others are major. One of these was Microsoft's decision in 2019 to discontinue security updates for their Windows 7 operating system. This resulted in large numbers of home users needing to upgrade to Windows 10, a shift that in my household necessitated the purchase of two new PCs, the jettisoning of one four-year-old machine and the trickle-down of another, slightly younger, to act as a non-networked word processing device. For households with gaming enthusiasts, hardware turnover is even more frequent, for while the consoles still function, game developers develop the latest games on and for the most recent platforms. Further, they imagine gamers as generational communities of practice. So, for parents of teenage boys in particular, it is almost impossible to avoid a commitment to either Xbox or PlayStation and to buying into the upgrades to these systems at the time of their roll-out. Not to purchase PlayStation 5, for example, would be not only to deny your child the chance of developing their skills on the latest games, but also of the capacity to play with and interact with their friends on that platform. When sociality is intensely technologically mediated, the effect of refusing to upgrade is potential social exclusion. What might look, then, like the height of a rampant materialism, when parents rush to buy the latest upgrade for Christmas, is actually more appropriately seen as doing parental love through consumption and in conditions of planned obsolescence.

Exactly the same sensibilities are exploited around a rather different category of consumer good: football shirts. The annual change of football strip by the top football clubs may be a branding necessity for them but it is also a major revenue stream, for it ensures that the majority of younger fans (in actuality mostly their parents, since they are the primary purchasers of such shirts) display their loyalty to the team by forking out for the new kit. In so doing, last year's shirt is declared by club and fans to be of lesser value – it's last season, 'ancient history' in footballing terms, kit worn by players who may even have left the club. Indeed, in circumstances where the currency of fan talk is always the last game, the next match, the current team, the relative merits of its players and the current league position, last year's shirt can only really be of commemorative value – and only really if the season was a memorable one. Often, especially for die-hard fans of clubs that habituate the relegation zone battles season after season, it's a season to forget. To continue to wear last year's shirt, then, is a disruptive act of not really fitting in. It may even be a sign of ambivalent loyalties – to individuals wearing a number, for example, rather than a demonstration of a more appropriate tribal loyalty to the team. And while that ambivalent position might be entirely reasonable for an adult fan to occupy, it is much harder for younger fans, for whom the need to fit in with the current team is overriding. So, at Christmas, alongside the latest Xbox or PlayStation gifts, goes the latest team shirt, if the child is a football fan. This version of parental love is enacted in conditions of manufactured obsolescence. However, this is not the manufactured obsolescence that we saw with consumer durables. Instead, it is more widely illustrative of the ways in which the manufacture and purchase of the new is hard-wired into the fashion system. Football shirts, then, act as a bridge to the final category of consumer goods that I want to consider here: clothing.

Not all clothing is fashion, but fashion is the primary driver of the apparel industry, which is a low-margin business in which the high turnover and replacement of goods is critical to continued profitability. In this regard, football shirts are well off the pace of what is the norm in the fashion business, where the gap between 'on trend' and purchase has now diminished to the point where the act of purchase marks the point at which a garment falls out of fashion. What once was an industry in which newness and its perceived desirability was choreographed by mimicking and foreshadowing the seasons (in which autumn/winter clothing went on sale in summer, and so on) has become an industry in which the mass market is defined by 'fast fashion'. This is manufactured obsolescence 'on speed'. It has not only narrowed the gap between the manufacture of a garment and its obsolescence as a desirable good; it also bakes disposability into that garment, for if it is out of fashion at the time of its purchase, then this pretty much ensures that the garment will be worn just once by the consumer, and then discarded, to be replaced by the new trend coming in to stores, or online fashion sellers, through the supply chain, and the next and the next and the next.

Disposability in itself is not a new attribute in terms of apparel. The first example of this is 'nylons'.[24] Manufactured by DuPont, nylon was in the vanguard of the synthetics revolution in fashion, and its first widespread application was in stockings. The trend for wearing stockings is connected strongly to the parallel one of rising hemlines in women's fashion. Nylons first replaced silk stockings in 1938 in the US, and by 1941, DuPont had captured 30 per cent of the US hosiery market[25], with the average woman buying eight pairs of stockings a year. In the immediate aftermath of World War II, such was demand that history records 'nylon riots'. In Pittsburgh in 1946, for example, a reported 40,000 women queued, desperate to get their hands on one of the 13,000 pairs available, and so much was demand in excess of supply that goods had to be paid for in advance. That situation, however, was as much an effect of the disposability of the good as of their perceived desirability. Albeit that they were advertised as the means to achieving sheer glamour and great legs, the reality of real women's legs and hosiery is the altogether less glamorous look of 'snags' and 'runs', which have the more than annoying habit of materializing at the moment one puts the things on, or pulls them up having been to the toilet. That known reality underpinned the purchasing strategies of me and my female friends from the outset of when we started buying hosiery in the 1970s. Hosiery was something that had to be bought in bulk and at multiple points in the year, in the full knowledge that at least some pairs would end up in the bin without even making it to wearing them once. Other pairs might be slightly more robust and last the day without developing these 'runs', but one could not rely on this, so days marked by important occasions, where look mattered, would see spare pairs of tights carried around in bags on a 'just in case' basis. In other words, hosiery required consumers to carry stocks. Moreover, beyond the challenges of everyday wearing, these are not garments that withstand multiple washes and wears. So, on average, most such purchases were discarded in under a month. We see here, then, with hosiery, the beginnings of the strong association between women's fashion, manufactured obsolescence and disposability, in which disposability is not only manufactured into the good but also informs a bulk purchasing strategy that may appear consumerist but that is necessitated by knowingly having to buy built-in failure.

That same relationship is replicated across many types of fashion good, if not quite with the same degree of intensity. Jeans, t-shirts, shirts and tops are all categories of clothing that quickly develop signs of too much wear, and often a degree of failure. In part, this is an inevitable consequence of wear. The stretching, pulling and movement that is what happens to clothing on the human body results inevitably in weaknesses at points of maximum strain, typically seams. Economical, least-cost, often offshore manufacturing processes render those key points of garment assembly weak to start with, as little excess material is allowed for seams, or the cheapest zips are used. And then there is the wear that comes from washing garments. So, manufactured obsolescence and disposability are built into much clothing, without even the dynamic of fashion trends. As we have seen, however, fast fashion has taken this to new extremes, switching mass market fashion's temporal dynamics from those that still govern the designer-led fashion industry and its annual cycle of runway shows, to those of the just-in-time supply chain, in which fashion is held in the supply chain and changes at speeds that can be managed within that supply chain. The effect in terms of clothing discard is enormous. In China, for example, an estimated 26 million tonnes of clothing is now being thrown away per annum, with less than 1 per cent of this being recycled.[26]

The pervasiveness of manufactured and planned obsolescence within contemporary capitalist economies ensures that consumer discard is baked in to an increasing number of economies. A consequence, especially in Europe, has been heightened debate over the environmental unsustainability of these forms of obsolescence, and there has been renewed interest in the possibilities of repair as a means to more sustainable resource use. The argument here is a compelling one, which seeks to use repair to extend the utility of things. In Europe, those concerns have recently brought about demands from policymakers for manufacturers to incorporate a greater element of reparability into their goods, especially electrical and electronic goods. Repair is seen to be the means to attenuating the social lives of these things and of avoiding, or at least delaying, what is seen as an oncoming tsunami of e-waste.

The final section of this chapter turns to consider these possibilities. It does so not from the position of either a trumpeting of or nostalgia for repair, but rather in full recognition of the challenges and constraints that consumer-heavy economies pose for the practice of repair. As I show, as well as raising questions of economics, much of this turns on how repair relates to practices of consumption.

On the limits of repair

Manufactured obsolescence, especially when allied to disposability, is the main current economic barrier to repair, but the economics of repair are also strongly connected to the value of a good: the higher the value of a good, the more likely that repair is built into its consumption in the sense of the use of the good. This is because the cost of repair continues to be considerably less than the cost of the replacement good. Perhaps the best two examples of what we might call the 'repair rule' are houses and cars.[27] Houses, particularly in countries where the price of land is high, where demand for housing outstrips supply and where property commands high prices relative to wages, are continually being repaired.[28] Veritable armies of small businesses specializing in multiple trades – roofing, general building work and maintenance, plumbing, heating and electrical systems, glaziers, joiners – are engaged in this work. Cars don't have the same degree of longevity as houses, but nonetheless, they are serviced at minimum annually and again, there is an array of small and large businesses devoted to their upkeep, from small general repair garages, through specialists in tyres, batteries and exhausts, to main dealerships. If cars are serviced by a garage linked to a dealership, they will have parts replaced to a schedule that reflects the anticipated wear and tear on a car in regular use and being driven an average annual distance. However, the cost of those replacements gets higher and higher once the car is beyond the warranty period, with more and more 'big parts' needing scheduled replacement year-on-year, and/or by a particular mileage. This, of course, is the strategy car manufacturers use to encourage replacement purchasing on a three-year cycle. It ensures both a steady flow of new car sales/leases and a supply of good used cars for the strong secondhand market.

Both houses and cars illustrate the connection between repair, demand for secondhand or used goods, and the relative strength of those markets for particular types of goods. In the cases of houses and cars, demand for these goods is high, and there is little by way of stigma attached to these goods. Indeed, if we think about houses, quite the converse: in the UK, old houses, most especially spacious older properties in prime locations, often command premium prices in the housing market, certainly well above new builds. A more open question is how these connections play out in relation to lower value goods, such as consumer durables. But this is exactly the experiment that European policymakers have just embarked on in countering planned and manufactured obsolescence by insisting that manufacturers increase the repairability of their goods.

The gamble here is on consumer demand for the option of repair, for the success of this policy intervention rests on consumers seeing repair as a preferable alternative to replacement. That may well be the case with consumers motivated strongly by sustainability and environmental concerns.

But, when we look at what consumers actually do, rather than what they tell us in surveys, they are the minority of consumers. Even with a high-value good such as cars, repair, for many, is at best an option with a limited life – something a consumer will do until the point at which replacement is considered the better (usually more financially cost-effective) option. With consumer durables, we can anticipate that, as with car manufacturers, firms will ensure that the replacement option will remain a sensible option for the majority of households. As with cars, then, increasing repairability may very well have the effect of increasing supply into the secondhand consumer durables market.

Another – and open – question is just how much demand exists in this market in European countries. Outside of the antiques and collectables market, demand for refurbished consumer goods currently is heavily skewed to lower income groups – effectively, the lowest quintile by income. Even allowing for further growth in inequality as a result of further fiscal austerity probably would not see demand expand much further. So, there are very real questions here that point to repair's contingency on economies. In conditions of relative affluence, repair is only really an option for the highest value goods. By contrast, in economies where average incomes are low and the relative cost of consumer durables high, the market for repaired and refurbished goods is far more extensive. This is why there are vast markets dedicated to refurbished goods in the global South. The likely effect of such a policy in economies where most consumers are relatively affluent is that this will saturate the secondhand market. What happens then?

Well, if other secondhand markets are a guide, the first thing is that increasing supply leads to finer category distinctions in the market, with demand increasing for the perceived higher quality refurbished goods that enter that market. This is where brand matters. In these circumstances, much as with the secondhand car market, one might envisage some consumers opting to buy a secondhand Miele appliance say, rather than a new Hotpoint, much as they opt to buy secondhand Mercedes and BMWs. The knock-on effect of that, though, is that demand for poorer-quality secondhand goods (cheaper brands, goods exhibiting more visible levels of wear and tear) falls off. What then happens in other secondhand markets is that these goods are only held for a minimal time in the secondhand market before they are sold into the scrap market, where they are of value only for their recoverable resources. The economics of repair versus replacement in consumer-heavy economies, then, may not only accelerate the passage of more consumer durable goods into the secondhand market, but also accelerate its passage to the scrap market. Ironically, rather than slowing down the consumption of consumer durables and stretching out the use of these things, insisting on their repairability may end up speeding up the passage of many of these things to their destruction.

It is just as important to acknowledge that repair is not entirely a matter of economics and economies. It is also about the relationship of goods to what the social scientific literature refers to as the 'practices of consumption'. In this literature, unlike economics – where rational choice purchasing behaviour is all that matters – consumption is not just about purchasing things, or their acquisition; it is also about using things to do other things. Using things is intrinsically bound up in what social scientists call their materiality. There are three dimensions to this that matter for our purposes here.

The first involves using things to do other activities or practices. An example illustrates the point. I spend a lot of time practising and playing my fiddle. To do that, obviously I need a fiddle and a bow, but I also need rosin, strings, spare strings, a shoulder rest, a spare bow, a cloth duster, a string tuner, a violin case, two music stands – one for practice and one for taking out and about – and vast stacks of sheet music. This is the array of things that make up fiddling as a material practice and that any competent fiddler needs in order to participate in this practice and to reproduce it as a practice. In other words, and in general terms, an array of related things stabilize a practice – they hold it together such that the practice can itself be performed and reproduced. This goes on across a plethora of activities – cooking, fishing, cycling, skiing, surfing, painting, photography and climbing, for example. There are literally hundreds more examples one could think of. As the literature on practices of consumption shows, much of our consumption of goods is orchestrated less by the desire for and acquisition of a particular good (the classic materialist impulse) and rather more by how these acquisitions relate to doing or participating in certain activities.[29] So much so that the sociological literature understands consumption as a meta-practice, with these sorts of activities being examples of constituent practices of the meta-practice of consumption.[30]

The second dimension is that these practices of consumption involve using things up. So, just to give a few examples, in the course of fiddling I routinely need to replace strings and have my bow rehaired, and playing a lot means that I use up rosin, which eventually needs to be replaced. My being a keen cyclist results in occasional punctures, the deployment of puncture repairs and, after two or three patches, the replacement of an inner tube. Less frequently it involves replacing tyres, brake cables, the occasional chain and front and rear gear shifters. Similarly, as an active hill walker, I get through walking boots pretty quickly. As an aside, all of this stuff ends up in the bin. I would hazard a guess that I am probably not alone in doing this with this sort of stuff.

The third dimension is a related one: the inexorability of material decay. Everything that has ever lived or been made and will be made is subject to this. Sometimes that deterioration can be sudden, as when a thorn spikes through a bike tyre, puncturing the inner tube, or when a violin's bridge collapses. More commonly, it is a gradual deterioration, as happens when rosin clogs up strings, dulling their resonance to the point where they need to be replaced. The wider point is that material decay or deterioration relates to an individual's performance of and capacity to participate in a practice. So, to continue the fiddling example, while it is perfectly possible to play a fiddle using 'dull' strings, the sound they produce is far from satisfying, meaning that they not only diminish the sound that results but also impinge on practitioner competences, and perceived competences. Dull strings diminish any player, of whatever ability; in social scientific terms, they destabilize the performance of a practice. If the practice is to be restabilized, then, they must be replaced.

As we will see, these three dimensions are particularly pertinent to thinking about repair.

Let's begin by going back to the consumer durables, home entertainment goods and information and communication technologies discussed in the previous section. All these provide good examples of how things in use connect to practices of consumption. Washing machines are integral to doing the laundry; fridge-freezers to storing stocks of food; Xbox to gaming; TVs to watching broadcast media and entertainment.[31] This is important because it means that when what we might term a 'key device' in a practice fails, then so too does the practice.[32] When the washing machine stops working, the laundry stops getting done – unless someone resorts to handwashing. The effect, very quickly, is a massive accumulation of unwashed clothing and household members minus sufficient clothing to wear (without buying more). When the fridge-freezer fails, not only are large stocks of food condemned to food waste, the capacities to replace and store those stocks are no longer available, which makes feeding a family challenging. It's a similar case with cars. If the car is in the garage being serviced or repaired, we might – if we are very lucky – be provided with a courtesy car, but mostly we are rendered carless – reliant on others to drive us about, on public transport or an alternative, such as a bike or walking.

The unravelling of established, taken-for-granted, habitual ways of doing practices (the laundry, food storage, getting about from A to B) matters when thinking about repair. To repair such a consumer good often involves removing it from its location in the home and taking it to, or sending it to, a repair premises. Sometimes the repair might be enacted in situ but the critical issue is not really the site of repair; rather, it is that it involves being without the device for the period of time that it takes to repair it. Depending on the good, this could take a few days or weeks. That time without is the period of time when a practice of consumption gets suspended. Depending on the practice and its importance in everyday life, such absences can be profoundly problematic. Being without a washing machine in my household for a week, for example, might not be a big deal, but for a household with young children in it who are in nursery and who need multiple spares of clean clothing provided daily, this is a major issue. It is a similar situation with devices like fridge-freezers. In such circumstances, replacing the good with a new version that does just the same function (possibly better) is not just cost-effective economics, it is also the means to the swifter restoration of the practice. For busy households, this is often the overriding concern. Intense and tightly coordinated household schedules depend on these consumption practices working on-demand.[33] Repair is not only a matter of being without capacities; it's also about its knock-on effects on schedule coordination. The more convenient alternative to repair is replacement.

Consumer durables and cars highlight that repair is often about the temporary suspension of a practice. But what about the material quality of their repair, if this happens? Both these categories of goods are ones where repair is a matter of substituting replacement new parts for the defective old part/s. The capacities and qualities of these goods (and therefore the practice that they enable) are unaffected by being repaired. Repair makes them 'as good as new'. Although we should note, in passing, that it certainly does not eliminate discard and discarding, which is what happens with many of the old, defective, worn parts that have been substituted. There are other goods to which the same principles apply. The bicycle is perhaps the classic example, much lauded in sustainability circles. Here pretty much every part can be replaced, and the bike itself can be rebuilt from the frame up, should one wish.[34] Similarly, laptops and personal computers can be refurbished and reassembled and work 'as new' or better.

Other goods are more challenging to repair, though. This is especially so when repair goes beyond a part or component and where it is instead an attempt to arrest rather more fundamental damage to, or decay of, an object. This kind of damage is often a direct result of wear and tear, or simply being out 'in the wild' of everyday life, full of other people, animals and interactions with other substances – dirty polluted air, rain and snow, mud, spilled food and drink, and so on. The best category example here is clothing. Clothing certainly can be repaired. I am reminded of a TV programme on curatorial skills that included as one of its examples the repair and restoration enacted by the textiles team at the V&A Museum in London on a Japanese samurai cloak that had been damaged by a sword blade.[35] Such skills pertain to the restoration of heritage value. They are about the rekindling by experts of one-off, seen to be significant historical objects of clear provenance. So, once they are repaired, they are put on display, typically protected by a curatorial regime of glass cases, cabinets, specialist lighting, carefully controlled temperature and humidity levels, and distance from any form of direct human touch. By contrast, there is no team of experts to repair ordinary clothing. Mostly it's up to the skills – or not – of the wearer, and the repair has to take its chance back out in the wild. That means the repair can often be what is described as 'a bodge' or a 'quick fix' – something that will suffice for a very short while, to hold a garment together, but that fails miserably to restore it to its former state. The result is that the inadequacies of the repair propel the garment to the category of rubbish value. In such circumstances, it often becomes consumer discard and is replaced with another garment.

It has not always been like this with clothing. Back in the 1960s in the UK, ordinary clothing was routinely repaired. My mother certainly learnt the skills of clothing repair, for I can recall whole evenings as a child where, after having cooked the evening meal, she 'turned' the worn collars of my dad's work shirts and mended ripped sheets, by stitching them together – although she drew the line at darning socks. Those repair skills had been taught to her by her grandmother, but the skills were not passed on to me. The reasoning, I later learnt, was a decision grounded in values: unlike cooking, say, or knitting (both of which domestic craft skills I was taught), clothing and textiles repair was no longer seen as a domestic skill necessary to be passed on to daughters. More fundamentally, it was also seen as a skill that carried normative conventions about women's lives, which defined them by and limited them to the domestic sphere. In the 1970s, the world was changing. Women had more important and self-fulfilling things to do with their time than mending clothes (like work and other leisure activities); the synthetic-based clothing that was starting to prevail by then was seen to be of insufficient quality to be worth repairing and more difficult or even impossible to repair compared with natural fibres; and, just as tellingly, who wanted to wear repaired clothes when they could be replaced cheaply by new ones? So, clothing repair was a site of women's resistance. None of my school friends were taught it, and neither did the all-girls school I went to see fit to teach it to its pupils. Instead, our class was taught sewing skills – first by making plimsoll bags and aprons in afternoons of total silence, on pain of class detention, while the teacher mended her husband's shirts and trousers in front of us! The effect has been not just the disappearance of those clothing repair skills but also the disappearance from homes of the apparatus of clothing repair – an array of different needles, multicoloured threads and leftover wool strands, thimbles, an 'oddments' (or 'remnants') box containing different colours, patterns and types of material, patches, and collections of buttons, press studs and old zips.[36]

This last point establishes a final point about repair and its relation to practices of consumption. This is that with the disappearance of a practice goes the discarding of the apparatus of that practice. This happens at the individual level, for example when people give up doing particular hobbies and leisure pursuits, for whatever reason. They then discard the apparatus of the practice. It can also happen when people become more skilled or more competent at particular activities and 'upgrade' how they participate in a particular practice – photography say, or cycling – switching from 'entry-level' equipment to more advanced kit. The domestic clothing repair example, though, is discarding at the societal level; it is the wholesale abandonment of a practice by a society. Such instances happen only when there is sweeping societal change, as happened here with the conjunction of feminism and women's entry into the labour market. Women's lives had changed so profoundly that they had neither the time nor the inclination to spend their evenings repairing clothes. Once change happens on this scale, it is difficult for the practice to be rekindled, for along with the loss of skills and competences is a loss of the tools that are necessary to be a competent repair practitioner. Instead, the practice becomes a 'lost art' – a way of doing things that has to be rediscovered and learnt anew.[37]

In contemporary capitalist economies, particularly those that are consumption heavy, repair as a general practice has become something of a lost art – for all the reasons this chapter has explored. For very good economic reasons, it persists in relation to high-value goods. It also exists in the interstices and gaps in the mainstream economy, often as part of the social economy, providing refurbished goods for people on low incomes and/or by providing sustainable alternatives to the 'acquire-use-discard-replace' cycle. But such is the reality of everyday life for most households that in seeking to attenuate the consumption of objects and things, the practice of repair all too frequently comes up short. This is not because of the failings or shortcomings of the act of repair itself; rather, it is because the need for repair marks when things fall out of social life. Repair, then, is disruptive of many of the very practices of consumption that striate and shape modern social life and on which consumer-heavy economies rely.

4

Conduits, Value Regimes and Valuation

Or, following consumers' discarded things

Chapters 2 and 3 have shown that discarding and consumer discard are inevitable facets of social life and the organization of contemporary capitalist economies. They have also established the insights that come from thinking about consumption not simply as the stuff that we buy, or even dispose of, but as stuff we do things with, for it is this that connects acquisition with divestment. One of the many advantages of this perspective is that it emphasizes that consumption – as this is lived by people in an everyday sense – involves multiple, synchronous but, for individuals, mostly sequential practices. If I am cooking the dinner, for example, I'm in the kitchen, working with organic stuff, turning food into a meal. Doing that also involves pots and pans, cooking implements, a hob and/or oven, maybe a recipe. I may also be listening to the radio or to a podcast, but while I'm doing this I most certainly am not driving the car, reading a book, playing my fiddle or out on my bike. The things that I need to do those practices, then, lie idle while I'm doing the cooking – and for much of the time, what we can think of as the social lives of things are like this. We actually use our possessions for quite limited amounts of time. The idleness of many consumer things has been one of the primary motivations behind efforts to grow what is termed the 'sharing economy', grounded in access to goods, through renting and leasing.[1] The way I prefer to think about idleness, however, starts from the reality that consumption for the majority of people is mostly still grounded in the ownership of many things.[2] This means that our things continually move in and out of the category that is the surplus.

The surplus as this relates to consumer goods works on a spectrum of intermittency, with different things spending more or less time in this category. At one end lie the things that we utilize on a daily basis, or multiple times a day – clothes (but not all clothes), personal hygiene products like toothbrushes and razors, food stocks, cutlery, plates and bowls, mugs, cooking implements, mobile phones and computers, often cars. At the other end are things seldom used – some items of jewellery, certain shoes, a particular suit – or things that are used once a year. Christmas decorations are one of the best examples of this. Somewhere in the middle would be things used on a seasonal and/or monthly basis. In the UK, lawnmowers and much gardening equipment lie idle for at least four months in every year; barbecues and garden furniture are in use for just the summer months. Rather nearer to the daily end of the spectrum would be the things we use for our key hobbies and leisure activities: walking boots, walking poles, outdoor gear, golf clubs, bikes and so on.

The surplus as a category also has a dynamic of its own – the capacity to begin to separate people from their possessions and to open up routes to what is known as divestment, or more colloquially, getting rid of our things or discarding. This happens when things come to be seen as superfluous, as no longer needed. These kinds of things emerge in multiple ways in the course of everyday life. Changes in living circumstances such as moving house are among the main triggers. Surplus things also emerge in the course of the practices that shape everyday consumption, as we saw, for example, in the previous chapter with getting rid of entry-level equipment in order to trade up to something that allows for doing a practice more competently, for example, a bike, new camera kit, skis, a better musical instrument.

One of the major transformations to have occurred over recent years is in attitudes to the surplus and to surplus things. For many people of my generation, faced on the death of their parents with dealing with their parents' effects, there is a general sense that theirs was a generation that held on to things, that never threw anything out because 'it might come in useful some time'. Theirs was a surplus borne of wartime austerity; a habit that meant things were kept to hand, in areas of storage (cupboards, lofts, sheds and garages) and kept accumulating. By contrast, for younger generations, used to the on-demand availability of goods and often living in smaller spaces, surplus things are not things to be held on to but rather things to be got rid of. The surplus is stuff that is to be discarded. But once something enters the category 'discard', what then happens to discard? How is discard itself organized? And how does this organization connect to waste? These are the concerns of this chapter. To begin to open this up, there is a need to explain two key concepts: value regimes and valuation.

A brief note on value regimes and valuation

In general terms, the social science academic literature that I draw on here understands value regimes to involve the use of categories of material objects in the construction of culturally specific notions of worth.[3] This short, and quite probably opaque, sentence – like much academic writing – is shorthand. It assumes an academic audience that is immersed in and thoroughly familiar with its terms of reference. It is also positioning, in that it signals a stance to particular topics of debate. So, in translation, what does this sentence really mean? What does it tell us?

One of the most enduring of social scientific debates, at the most abstract level, concerns the relationship between humans and the material world, as in what we think of as the natural environment and our fabricated, and often manufactured, environment of objects and things. Within that, an unavoidable and huge strand of work in political economy starts from the foundational texts of Marx and, in particular, his account of the commodity, or things that are made and then exchanged for money. So influential did this account become in much of the social sciences that things came to be thought of as if they were only commodities, that is, the only things that mattered was the category 'commodities'. More than this, the value of things came to be defined narrowly, in relation to the manufacture of things and their exchange for money. So, when we have a statement that includes categories of material objects and a broad definition of value that is labelled as worth, we can see immediately that the relationship between humans and the fabricated world of objects and things is being seen in terms that go well beyond commodities and their exchange. Hence (and to complete the task of unpacking the sentence), Appadurai's opening salvo in *The Social Life of Things*: 'A commodity is not a thing – it is a thing in context'. In such a way, the pathway is open to thinking in terms of the social lives and biographies of things. Things may start off as commodities but in their social lives, they move out of commodity exchanges, perhaps to become ordinary things in everyday consumption, such as the mug in front of me as I type, which is one of several that I've had for at least 20 years. Alternatively, they can become a very different category of things: objects that do memory work. These types of things – often they are in the category of gifts, souvenirs and mementos – are endowed with memories of significant people, places and times. A good example of such an object would be rings that are passed down from grandparents to granddaughters, mothers to daughters, aunts to nieces, through generations of a family. But these types of objects can also be souvenirs and mementos of places – a category of things that has become of particular importance in recent years with the rise of gap years and international tourist travel.[4]

Objects that do memory work are in a different value regime to the mug. Although pretty much any object can work as a memory object, it is the particularity, or singularity, of the object that matters here, that is, their worth to individuals; no other object can do the work of rekindling particular memories for that person, so no other object has the capacity to substitute for it. Equally, that same object, by definition, cannot work in the same way for another person, precisely because the attachment, forged by memory, is absent. By contrast, the mug's worth is typical of many things in everyday consumption. It combines utility and aesthetic, but it can be replaced should it break. In addition, that same mug could be of the same worth for another person, precisely because worth here lies in a combination of utility and aesthetic. Immediately, then, we have two value regimes comprised of categories of material objects, where the context is defining their worth: one is that of things in everyday consumption, the other is things doing memory work. A third value regime comprises all those things that have been manufactured or assembled and that are now for sale but at this stage are purely commodities. These are goods whose economic value is captured in a price but whose worth to their seller depends on their sale.

Let's turn next to valuation. The process of valuation is the precursor to the creation of economic value as well as worth in its more general sense. Valuation processes are often rule-bound but always context specific. Typically, they entail the interplay of two or more political, economic, material-technical, social and cultural factors, but how that interplay works depends on the nature of the category of the good, those who do the work of valuation and, in market societies, the specificities and inter-relations of particular markets. So the same good can be open to varying valuations and if it is offered for sale, realize more or less value.

As an example of valuation and its relation to value in a market setting, take the sale of tens of thousands of yearling[5] thoroughbred horses each autumn. All young thoroughbred horses are assessed on the basis of: 1) their pedigree (or paper provenance) value, comprising the commerciality of the sire (father) and his stock and the racing and breeding record of the dam (mother) and related family members over three generations; 2) physical type (size, scope, general make and shape, conformation – faults, and so on, and movement/paces); and 3) results from veterinary inspections. Assessment, therefore, involves the interplay of material-technical, economic and social/cultural factors, as encoded in the history of horse racing and the record of the results of racing thoroughbreds over centuries. Assessment defines the worth of any one horse, initially to its owner/breeder. However, if the horse is to be offered for sale at the big yearling sales, as most are, it will also be assessed by representatives from one or more of the sales auction houses. In this case, the outcome of their assessment is not just a valuation (which may, or may not, accord with that of the owner/breeder) but also a classification. The classification positions the horse in a wider field of anticipated value, defined by the yearling crop that is being offered at auction. Sales are stratified. The premier sales are reserved for those horses considered by the sales houses to be of the most commercial value. This means that only a select few hundred horses are accepted into the flagship sales, which attract the biggest spenders and command the highest prices. The next most prestigious sale takes the next slice of the crop by valuation, and so on down the hierarchy. But even within sales there are gradations of value, with particular days reserved for the most highly valued individuals in any one sale and specific times of day allocated to what the sales house considers to be the cream of each sale. The choice facing a vendor, then, is whether or not to accept the sales company's classificatory value of a particular horse (captured by a sale and a lot number), for that will have major ramifications on the financial return. In such a way, we see how the worth of young racehorses (the material category) is constructed, and within that how the worth of each individual is estimated (through a process of valuation) and then realized as financial value through the auction itself.

★ ★ ★

How does all this relate to discard? Since discard is frequently defined as waste and waste is defined as discard, it is tempting at first sight to think of discard as a homogenous material category – as stuff of rubbish value (Chapter 1). As this chapter shows, this is an oversimplification, and misleading, because it neglects two issues: 1) the question of how that part of the surplus that is identified as open for discard intersects with value regimes; and 2) consumer processes of valuation.

There are three primary ways by which consumers realize worth from objects that enter the category of the surplus that is to be discarded. The surplus can: 1) be given away to others to use (effectively, gifted); 2) re-enter the commodity phase by being offered for sale, or used to realize money in the form of credit; or 3) be discarded as rubbish value, be that through the medium of the bins and containers associated with waste management in whatever form that takes in any one country or municipality, or via open dumping, as with 'fly tipping' or the discarding into open drains that occurs in informal settlements the world over.

These three routes organize and choreograph things differently. In that way they can be thought of as distinct regimes – or as systematic ways of organizing and effecting the handling and treatment of things, in this case in the course of their discarding. The three regimes span cultures and they co-exist, but their relative weight in different parts of the world varies, most notably with respect to the pervasiveness of the waste management industry. Within each regime, too, depending on the context, there is a greater or lesser range of possible routes, or conduits, available. So, the precise mix of what goes on within each regime will always be specific. Furthermore, the route any one discarded object takes is dependent upon the valuation of it by the consumer in whose possession it has been. That valuation affects the conduits, or channels, in which a thing is placed and the regime into which it initially passes. It is critical to what surplus things become, for conduits shape and mould not only these but also future possibilities, opening up some routes while foreclosing others.

Mindful of the highly context-specific nature of the value regimes shaping consumer discard and its valuation, in the remainder of this chapter the focus shifts exclusively to the UK. The three primary regimes shaping discard here co-exist but, as I show, how they each work puts a different premium on the relationship between economic, material-technical, political, social and cultural factors. That, in turn, entwines with consumers' own value systems and their valuation of the things they are discarding. Broadly speaking, however, people living in the UK can and do go to considerable lengths to endeavour to ensure that the surplus things they are discarding do not automatically end up as stuff of rubbish value.[6]

Discard's three regimes of value

The first regime, of giving things away, relates closely to the social economy. Here, the worth of activities is defined in terms of what they achieve socially.[7] The defining characteristic of things entering the social economy is that, whatever the thing, it is given away in and of itself, without any monetary transaction. Those who benefit from such acts often realize money from their sale, as happens when unwanted furniture is given to a furniture reuse social enterprise, or when no longer wanted clothes are given away to charities. But, in the social economy, consumers act as donors. Unlike financial donation, however, their relation to exchange in this case is unmediated by money; it is exclusively about physical goods. Altruism aside, this means that we need to look closely at what else is going on when people give away their things, for there is no such thing as a free gift. This, undoubtedly, is a highly contingent issue, for there are multiple ways in which things are given away in the UK, each with differing effects. Sometimes they are directly swapped; often they are given away through social networks – family, friends, community groups, neighbourhoods; but they can also be donated in circumstances where the social ties are shallower, where the recipient is an organization, an intermediary to an unknown other and a means to collective redistribution. All of these contexts shape the meaning of donation. However, collectively they highlight that the value being realized by participation as a donor of used goods in the social economy is not financial; rather, it is social and cultural. This is discard's worth to the original consumer here.

The second regime seeks to remonetize goods in consumption. Here, the goods in one's possession are seen to have a latent financial worth that can be turned to money. So discard's worth here is mostly financial. This has long been recognized by the urban poor. In 19th-century England, for example, the pawnbroker was the primary means to realizing cash, in the form of credit, from limited assets. The repetitive pawning of his coat by an impoverished Karl Marx in 19th-century London[8] is one of the most famous examples of this practice. The 'rag and bone man' was another to cash in on the latent financial value in goods. These itinerant traders were often 'travellers' and, even in the mid-20th century in the UK, they still called door to door to the call of, 'any old iron?' Iron was a euphemism, for the rag and bone man was a trader in most forms of scrap, particularly metal. They would pay a small amount of cash for unwanted pots and pans, mangles, fridges and washing machines, wrecked bike frames and the like.

The later decades of the 20th century and the early 21st have seen this regime diversify and intensify in the UK. The conduits for turning goods to money now encompass the face-to-face physical world and the digital world. Yard sales, garage sales and car boot sales, or table-top sales, are just some of the conduits available to consumers looking to turn their surplus goods back to money. But it is the proliferation of online secondhand marketplaces, notably eBay, Gumtree and Preloved[9], that has really revolutionized secondhand exchange, disrupting what were previously strongly localized markets for used goods and making them regional, national or even international. The ubiquity of online payment systems such as PayPal enables geographically distant consumer-to-consumer exchange and has worked to overturn the dominance of cash as the privileged monetary form in the secondhand marketplace. Through offering parallel 'new' and 'used' listings, the major online marketplaces (eBay, and now Amazon) present used goods as an equivalent alternative mode of provisioning, much as a garage sells both new and used cars on its forecourt. They therefore disrupt the stigma that is commonly found around secondhand goods.[10] But perhaps most importantly, the sheer volume of goods offered for sale on these platforms, and the ease of becoming an online seller, has encouraged more people to participate in online marketplaces as traders. Online trading has become seen as a means to make money and as a source of additional household income. The consequence is that consumer goods start to become seen as stocks. Rather than being valued primarily for their use, in this regime, consumer goods become a stock of financial value to be realized through trading.

The third and final regime orchestrating household discard is shaped by the waste management industry. Since the late 1990s in the UK, this has been reconfigured from a regime that managed wastes – in the main by disposing of them in landfills – to a regime that treats and recovers materials that have first been categorized as 'wastes'. This waste-to-resource regime applies mainly to the material remainder of consumption. This is all the stuff that is not given away for reuse and redistribution, or that does not enter secondhand markets. But, as any perusal of the material handled by the waste industry will show, included alongside the remainder of consumption is stuff that is recognizable as a consumer good. This in itself is an indication of the convenience that the waste management industry's infrastructure offers as a means to get rid of household discard. But what it also indicates is that what is deemed to be of rubbish value is itself a highly contingent valuation. It depends on who, living where, is making that valuation.

The chapter proceeds by delving deeper into each of these regimes in turn. It then highlights their relation, for while these are discrete regimes, it is important not to lose sight of the fact that stuff cascades between them. Stuff that is listed for sale on a used goods platform, for example, may not realize a bid and, as a consequence, end up in the residual waste bin. People may try, but fail, to give something away through social networks, or through an exchange platform and, as a result, the item gets taken to the local household waste and recycling centre. The same failures are to be found when divestment occurs through intermediaries. Stuff donated to a charity may not get sold and is then pushed in alternative directions – either to another outlet in the same chain or (if it is an item of clothing) into the recycling market.

We begin in the social economy, in which discarding becomes giving things away.

Discard as giving

Consumer discard enters the social economy in multiple ways but in all cases, and regardless of the conduit, this is a form of exchange minus money.

Non-monetarized exchange is the bedrock of economic anthropology, in which the exchange systems of indigenous peoples has been an enduring concern. Highly influential texts by Mauss and Malinowski laid the foundations to understanding what has become known as gift exchange.[11] In particular, they emphasized its relation to status, power and prestige; the importance of debts and the uneven and unequal capacities for reciprocity; and the aura of the gift. In so doing, they stressed that what matters in this kind of exchange is not the gift in and of itself, or what that gift actually is, but the social relations and particularly the ties, constituted by and through the exchange. The classic examples of gift exchange provide some elaboration.

A brief note on gift exchange

The anthropological literature is replete with examples of indigenous non-monetarized forms of exchange and its relation to gifts and gifting. No one example is the same and interpretations of what exactly is going on in these exchanges are messy and often highly contested. Nonetheless, the basic principles are relatively clear. These exchanges are a means to establish and display individual status, rank and social prestige, and to creating social ties, positive and negative. They also work collectively, as a means to display the wealth of a family, clan or tribal unit and to build political alliances. As such, these exchanges are highly competitive and often ruinous in that accumulation is central to successful participation.

The classic examples in the anthropological literature are the kula ring, potlatch and moka. The kula ring was first described by Malinowski, based on his work in the Trobriand Islands in the Western Pacific during World War I. At the heart of this is the exchange, by canoe, across hundreds of sea miles of shell necklaces and bracelets, with red shell necklaces circulating clockwise between island communities and white shell armbands circling anticlockwise. These things did not remain long in any one person's possession; instead they had to be passed on within a given time frame, but in ways that ensured they circulated back whence they had come. Kula exchange, therefore, is the exemplar case of inalienable possessions – or the paradox of keeping while giving.[12] These are gifts that need to return to the original giver. Potlatch, by contrast, is the term reserved for the gift giving feasts that are found among indigenous peoples living on the Pacific North-west coast of the US and Canada. To host a potlatch requires accumulating valuables, which are then given away in overt displays of wealth. Such events are linked to feasting and festivity and are often associated with key familial and clan events, for example, births, marriages, deaths, but they are not limited to these. The moka exchanges found in Papua New Guinea are status based. Originally based on the exchange of pigs, the moka is the term reserved for an increment – the bigger the increment, the more the status. This is encoded in the term 'Big Man' – a status that is overtly contrasted to the 'Rubbish Man' whose inability to match the size of the gift, or failure to repay it, accounts for their lowly status. The modern equivalent of moka exchange involves pigs, money and consumer goods. In the film *Ongka's Big Moka*, we see how Ongka accumulates the goods necessary to grow his reputation as a Big Man, and the prestige attached to himself and his tribe. In sum, this amounts to 600 pigs, A$10,000, 12 cassowaries, 8 cows, a motorbike and a pick-up truck.[13]

★ ★ ★

Consumer discard in the UK is not like this. For one, it is mostly not socially valuable. Then, when it is shed – or more colloquially, got rid of – often the most significant facet of the process is to lose its association with the previous owner. Perhaps the best indication of this is how purchases of secondhand clothing invariably are subjected to cleaning rituals before being worn – a process of washing and ironing that symbolically detaches them from previous wearers.[14] At the same time, consumer discard is more likely to involve objects that are of functional and/or aesthetic worth, where the connections between people and things are looser. These are possessions minus personhood. Such connections are even looser when discard is genuinely the remainder of consumption – leftover bits of wood, half-used tins of paint, a few bricks, for example. Nonetheless, it is still helpful to work with the insights from this anthropological literature when thinking about what is going on when consumer discard heads in the direction of the social economy. This is because when discard enters this value regime, ridding becomes giving. Inexorably, in relations that involve giving there are vestiges of gift exchange. How this works depends on the conduit.

What are these conduits? In the UK, the sites span a spectrum of non-monetarized exchange. At one end there is a pure exchange, where all participants in the exchange donate something to the exchange and where these donations are then exchanged as equivalents directly between participants. Swap events are the classic instance: women getting together to swap clothes[15] rather than buy new things; book-group members meeting to exchange books between themselves, rather than buying something new to read. At the other end of the spectrum are those sites where discard is given away for free with no expectation of any return exchange. Here, donated things are turned back into goods, which are then resold, with the financial proceeds going to the intermediary. Charity shops; charity door-step collections; the old-style jumble sale, or its modern equivalent, nearly-new or new-to-you sales; and furniture reuse networks are all examples. Then, and rather nearer swap events in character, there is the platform-mediated version of the social economy, in which the platform acts as the intermediary, linking those with stuff to give away with others who want stuff. The best known example of this is Freecycle – a global network of relatively local geographical groups, with members' listings itemized as 'offers' and 'wanteds'. A more recent development would be the food-sharing app, Olio.[16]

The language of donation and offers that is used in these sites is an immediate clue that the discard that travels in these directions and that passes through these conduits is thoroughly entangled with the workings of gift exchange. But how precisely does this work out? Three instances, each of them inflected through the concerns of the anthropological literature on exchange, help to illuminate. These are: baby clothes and baby things circulating in the hand-me-down-around economy, charity shops and charity bag drops, and Freecycle. Collectively, they show how discarding can become gifting, and its effects.

Circulating baby clothes and baby things – a modern-day kula ring?

Baby clothes, baby things and young children's clothing and toys are one of the biggest categories of goods circulating in secondhand exchange.[17] They can be found everywhere and in large quantities – in nearly-new sales dedicated to baby things[18], in charity shops and also in sites that turn consumer's discard directly to money, such as car boot sales and eBay. So, we can say that baby stuff as a material category has no necessary connection with any one of discard's value regimes. But one of the most interesting facets of this category of things is that this stuff also circulates pretty much invisibly outside of these sites, in and around groups of mostly middle-class mothers in what is known as the 'hand-me-down-around' economy.[19] Some of these groups consist of kin relatives, typically sisters[20]; others are larger and are friendship-based, or work-based, groups. Tellingly, they are also groups that can act 'at a distance', with stuff being saved to be handed over at meet-ups – in other words, they do not always depend on geographical proximity for the stuff to circulate, although this certainly does help.

The way in which these exchanges work is that the mother with the oldest child of either gender in the network acts as the primary giver, passing on stuff as and when it has been outgrown to the mother of the next oldest boy or girl in the network, and so on. However, these gifts are often inalienable gifts in Weiner's sense of that term, for the network works on the understanding that things given may well need to be returned to the original giver when the next sibling arrives. So, stuff stays within the network, at least until that mother decrees that that is the end of reproduction! At which point the stuff is free to circulate more widely. So it is that we find younger sisters holding on to prams and piles of baby clothes just in case an older sister 'has another one', only releasing them into the wider secondhand economy once that possibility is decreed to be no more.

One of the benefits of this kind of hand-me-down-around activity is that it means the group as a whole, or collectively, purchases far less by way of new goods. Undoubtedly, that is one of the primary motivations for participation. But is this the sum total of what is going on here? Probably not, for if we think with the anthropological literature on gift exchange, what becomes more prominent is the power in the network of the original mother, to whom all the other mothers are indebted (and who can only repay that debt by returning the gifts to her). Circulating baby clothes, here, have more than a whiff of kula exchange about them. Beyond that, such exchanges signal the importance of baby clothes for holding on to a sense of babyhood and to counter baby loss. For what is circulating here is X and Y's baby clothes, and in that circulation is the means to keeping the memory of that baby girl and boy strong, most especially as that child inexorably grows older and bigger. It is a means to using baby things to attend to, and counter, a mother's loss of babyhood. Keeping these things circulating is how mothers in general valorize babyhood. But, keeping these things circulating within social networks is a much more potent version of the same, for seeing another baby wearing those clothes is the means to rekindle memories of their child in baby form. This is one of the reasons why baby clothes have such an aura, why they are valued so much and why they circulate so strongly, even in advanced capitalist economies.

Charitable giving? Charity shops, bag drops and the debt of overconsumption

Consumer discard given freely to charities flows through two primary conduits: charity shops and charity door-step collection bags, or bag drops.[21] The emergence of the second of these conduits in recent decades highlights charities' recognition of the importance of convenience in capturing consumer discard. Mostly, this reflects the ongoing struggle charities face in securing (quality) donations.[22] That relates to the wider context. In part, this is about cheap manufacturing and fast fashion – much clothing is not designed to last. But it goes beyond this. As the conduits available for consumer discard in the UK have proliferated, so charities have had to work harder than they previously did to capture discard and to retain the volume of discard that they had grown accustomed to receiving. Bringing collection to consumers is one of the tactics adopted. Turning to convenience, however, flags that, actually, the specific charity isn't really that important to many of the consumers who discard stuff through these conduits. That finding emerged strongly in some of my previous research, where people would refer to dropping stuff off at the most convenient charity shop for them – typically the one with parking right outside the door, or by their bus stop.[23] Often they did so with seeming little knowledge of what the charity was, referring to it in only the most general of terms – 'the animal one', 'something to do with kids', and so on. So, when bags are posted through house doors by a charity, often on more than one occasion a year, and when we see that being done by multiple charities, we can see that charities are recognizing that sharp reality. The hope is that the prompt of the bag (hopefully their bag) will either nudge consumers to have a clear out or be a means to capturing stuff that has already been sorted but that has yet to make the journey to a charity shop.

There is more to these conduits than convenience, though, important though this undoubtedly is. Again, thinking with some of the central tenets of the anthropological literature on gift exchange helps to illuminate what is going on, and to show that the turn to bag drops may have unintended consequences for some consumers.

Charity shops work both to absorb a sense of overconsumption, as manifested by piles of surplus stuff hanging around a home, and, through their capacity to satisfy the bargain, to legitimate further acts of consumption. This is recognized in academic research on acquisition through charity shops, which identifies one of the attractions of buying in these shops as their appeal to the bargain, which in turn encourages more purchasing.[24] But charity shops' relation to overconsumption also works in another way – through their capacity to keep on absorbing donations. What is going on here, I would argue, is much as Parry argued with respect to charitable alms giving in

India, in which the pure gift of alms is seen to be poisonous. This is because the gift of alms embodies the sins of the giver, which then transfer to the priest, who is then saddled with these impurities, establishing a set of debts (or burdens) that cannot be dealt with. In the case of charity shops, the sin is consumption – and particularly overconsumption – while the role of the priest (in absolving the sin) is being taken by the charity. However, unlike the priest, charities are able to discharge those debts, through the sale of these donated goods. In sharp contrast, the charity bag drop seen through the lens of gift exchange comes extremely close to a form of begging by the charity concerned. Unlike charity shops, they are actively seeking donations; the bags are acts of solicitation. Of course, solicitations – much as with those of the beggar – can be ignored, passed by or overlooked (as when I put many such bags straight into the bin), but the effect on consumers of receiving repetitive drops I would argue is much like the beggar's continued presence. The endless drip-drip of solicitations provides a continual reminder of the inequities and inequalities that underpin contemporary consumption, while simultaneously suggesting that bags be filled and put out. What appears to be offering convenience might, then, appear more a 'poisonous gift' to consumers – one that in this case saddles consumers with a reminder of their continual consumption and the need for it. Perhaps that is why I put so many of these bags straight into the residual waste bin?

Freecycle – idealized communities and 'free' gifts

Founded in Arizona in the early 2000s and now a global network, with an estimated nine million members organized in well over 5,000 groups, Freecycle is distinctive from charity-mediated versions of the social economy in that its primary purpose is environmental stewardship: keeping stuff out of landfill. As such, it has attracted much attention in sustainability and green consumer circles, where its distributed model, volunteer-led, and locally based member networks are heralded and celebrated as illustrative of either, or both, a circular and sharing economy and as offering proof of an alternative collaborative model of consumption that counters capitalist-based consumerism and overconsumption. My own reading of this conduit is rather more circumspect and more in line with the altogether rarer critical appraisals in the academic literature.[25]

As well as being inflected through the 'social life of things' approach, my reading of Freecycle rests on what is known as a relational approach. While Freecycle certainly can be examined in isolation, as a distinctive site of secondhand exchange, one of the key points to understanding it is that it has emerged in, and overlays, an analogue mode of giving things away freely. The platform therefore sits in relation to other conduits: hand-me-down-around, the various charities' (or goodwill) sites, furniture reuse networks, and so on. While for some consumers, particularly the affluent and well-off who do more of their consumption online than offline, Freecycle may take precedence, for others there are gradations between what gets placed in various conduits – and what might be offered, for sale, on other used goods platforms. In other words, consumers' valuations do the work of placing goods in particular conduits. Typically, these valuations are grounded in the material qualities of items, most especially how used (or not) an item is and its brand.

One sees something of this at work in the listings of 'offers' on some Freecycle groups in the UK. Whereas there are, undoubtedly, many instances of goods offered that are in good condition, there are also many others where listings explicitly state that the item is broken, in need of 'TLC', thoroughly wrecked, and so on. This is the kind of stuff that does *not* appear in the hand-me-down-around economy or in charitable donations. Similarly, there are listings of oddments and leftovers, stuff like bricks, bits of carpet, shredded paper and planks of wood. Rather than take the 'taken' status listed against such items at face value and see this as automatically indicative of a community of active repairers, if we think through valuation we can see that such offers are indicative of goods being placed in a conduit that is perceived appropriate for broken and leftover, or remnant, things. For many consumers, then, this is a conduit that is closer to rubbish value than say a charity shop or a charity's bag drop, which is not to say that the things offered *are* of rubbish value, but rather that the valuation of whatever it is by their owner works to place them in a conduit that has stronger connections to rubbish value.

So what is going on when such items are 'taken'? It is tempting, particularly in an organization where environmental stewardship is the primary motivation, to see this as just that – as indicative of a commitment to rescuing, salvaging, refurbishing and generally reimaging old things. And, undoubtedly, some of this does go on; the platform can be used in this way. But, as the previous chapter showed, the capacities to live in this way are limited, at least for the majority. Repairing, rescuing and salvaging things are time-heavy commitments that often run with lifestyle changes, be they forced or chosen. Moreover, the openings and possibilities Freecycle's free pickings offer for networks of traders should not be underestimated. For every hobbyist repairer, there will be many, many more with direct or indirect connections to the extensive global trade in used and scrap goods. Rather than see Freecycle as an actual community committed to repair, reuse and refurbishment, then, I would see it more as an idealized community – one in which the size and scale of the network does the work of promotion of an imagined alternative future of repair, refurbishment, and so on, while decoupling that from the inevitable dependence of the network on inputs of manufactured consumer goods.

Then there is what to make of the used goods that are listed on the platform as in good condition. This takes us back to convenience but this time combined with the notion of singularization that comes from research on the social life of things. First, convenience: Freecycle may take some work on the part of consumers looking to get rid of stuff by giving it away – in that it requires joining a group, composing posts, sorting through emails from those seeking to acquire the good/s one has listed, and sifting through emails deciding who to give something to – but what the platform does is to remove the work of taking unwanted, surplus stuff somewhere else. Instead, the scheduling challenge of connecting unwanted stuff with potential beneficiaries is displaced onto the receiver, who typically has to collect the item/s. This means that the scheduling of this activity is controlled by the giver, not by the recipient or by the opening hours of a charity shop say, or the local household waste recycling centre, which might be alternative conduits for the item/s. So, an undoubted part of Freecycle's attraction is the convenience the conduit offers to those getting rid of things. Second, singularization: Freecycle involves gifts between strangers and is therefore totally different to what is going on with constituting gifts with known others, yet these exchanges are typically of individual items and they are asymmetric exchanges, in which the giver has the power to choose who they give to and from whom they withhold. Persons and who they are (or appear to be) really do matter here, as has been shown by research conducted in Freecycle groups in the north-east of the US. This identified the importance of selecting appropriate 'socially deserving' recipients for good-quality stuff and the types of screening mechanisms participants deploy in doing that.[26] Although these are gifts between strangers, identifying the right stranger to receive the gift really does matter. In such a way, the importance of personhood to possessions, and of attaching former possessions to the right kind of stranger, is demonstrated to be at the heart of many Freecycle exchanges.

In summary, then, and looking across these social economy conduits, we see an unavoidable truth: the used goods discarded in this way require those deemed to be 'socially deserving' to legitimate their ridding. In such a way, I would argue, the work of the social economy, and its worth for consumers, is for the sin of overconsumption to be passed on and absolved. It is this that allows for the guilt-free acquisition by the giver of replacement new goods. But when we think hard, and with the help of the anthropological literature, about what is really going on in many of these exchanges, we see that discarding in this way is also about passing on burdens. For what is happening here is the extension of the social life of things *combined with* the transference of the burden of their potential to become waste onto others. As with so many forms of gift exchange, then, in giving stuff away for free, a debt is established that cannot be erased. It is the recipients of these free gifts, then, who bear the burden of dealing with the physical effects of society's collective overconsumption. They are the modern equivalent of the Rubbish Man (sic), who face the impossible task of making free gifts from stuff that shows the material effects, in decay, of its continued consumption. With these sorts of things, there is only so much repair and refurbishment that can go on, or, said slightly differently: the social lives of things are not infinite. Instead, the hard reality is that consumption done through increasingly used things often involves dealing with a lot of overly used things that can quite quickly come to be seen as of rubbish value.[27]

Realizing latent financial value from possessions – from pawnbrokers to platforms

Since the birth of money, possessions have been a means to realizing money. There are two routes to this. One, through the intermediary of the pawnbroker, turns possessions to credit, through the medium of consumer debt. The second, through resale value, turns possessions to money in hand, or in the bank, with no liabilities attached. Although distinct, the two routes are connected by relations to resale markets. For pawnbrokers, the estimated resale value of an item, based on estimates of what it would realize in various markets, is the basis for calculating loans. It is one of the estimates used to underpin the loan (and interest) that is offered against a pledge and, for the pawnbroker, works as security against the loan. By contrast, when possessions are actually offered in used goods markets, rather than merely held against them, valuations become the basis for the potential transfer of property rights. Whether that happens is down to whether prospective buyers and sellers can translate valuations to a mutually agreeable price.

Pawning the surplus

Goods, or possessions, have long been a means to realizing credit, most especially for those without ready access to it. This is what explains the historical ubiquity of the pawnbroker across societies, be they Chinese, European, North American or South Asian. Historically, the pawnbroker (as opposed to the pure moneylender) has been the primary alternative to the bank. Pawnbrokers, then, are a means to access credit for those who are either excluded from the formal banking system or who prefer not to use banks, for whatever reason – and possessions are vital to accessing their credit services.

In Victorian England, the pawnbroker was integral to life for the working class, for whom pawning goods was frequently the means to paying for food, fuel for heating and housing rent. In the absence of regular wages and a welfare state, possessions were the means to access credit and to avoid the workhouse.[28] So, pawning was part of everyday finance and household reproduction[29] for the majority of people – and there were pawnshops on most urban streets.[30] Historical research gives an indication of the scale of the phenomenon. Evidence to the 1870 Parliamentary Select Committee suggested that some 207 million pledges were made annually, with 30–40 million of those being in London. By 1911, British pawnbrokers issued some 253 million loans, with an estimated 140 pawns per person per annum.[31] Mostly, these pledges were of low value, often of less than two shillings. This pattern – of the repetitive pawning of low-value possessions – says much about the nature of the goods being pawned. They were things people could do without in their daily lives, on a fairly frequent basis, not the things that people couldn't live without – that is, they were surplus things. So it is that seasonal clothing such as coats or weekly 'best' clothing, such as a Sunday suit, were pawned on multiple occasions in a year, whereas things like cooking pots were not.

In this way, we see how the notion of surplus goods works, even in poor households where consumption is mostly at the level of day-to-day survival. But what we also see with repetitive pawning over the course of a year is that goods can flow in and out of the category of the surplus on multiple occasions, without ever being divested. One of the advantages of the pawnbrokers for the urban poor of 19th-century England was precisely this: while pawning goods may have temporarily removed rights to the use of a particular good, provided the debt was repaid, it did not relinquish property rights. In such circumstances, the value of at least some household possessions lay less in their form (the coat, the Sunday suit) and rather more in their continued capacity to keep being turned into money.

At first sight, all this is rather different to the modern pawnshop, where the vast majority of pledges comprise jewellery, gold and watches. Mostly, the modern pawn business involves large cash loans secured against valuable, luxury items[32] – an apparent world away from the low-value loans secured against apparel that characterized the 19th-century equivalent. But is this really so very different? Industry sources[33] indicate that their services today are heavily used by minority households, most especially from cultures where, in addition to an avoidance of western banking systems, there is a strong tradition of buying, selling and pawning gold, and where dowries are a liquid form of wealth, typically used to finance major events in the reproduction of the household – marriages (and funerals), healthcare, school fees.[34] Although the goods that secure the loan are different, their value – as surplus goods that are the means to financing household social reproduction – is much the same. The only real difference is in the capacity of these goods to realize money. For jewellery, gold and watches, unlike apparel, are stocks of wealth, possessions accumulated precisely for their capacity to become liquid money.

Thinking through stocks and money is also increasingly important to understanding what is going on when consumers discard goods by offering them for sale in offline and online marketplaces.

Surplus goods, stocks and money

Exchanging surplus goods for cash has a long history in England but the last 30-odd years has seen some important changes. At the turn of the 1990s, when I first moved into the village where I live, there was a notice board in the Post Office. The board functioned as the means to informal intra-community exchange, invariably conducted through cash. Notices of things for sale (furniture, bikes, children's swings and trampolines, garden equipment such as lawnmowers, hoses, and so on, even cars) would routinely appear there, alongside items wanted and services offered and wanted, such as cleaning, childminding, gardening and mobile hairstylists. The board has long since disappeared – a casualty of Post Office closures and then of the transition of much community-based exchange to local Facebook and WhatsApp groups. Rather more extensive in geographical reach than local notice boards were the classified sections of local newspapers[35], in which sections comprising categories of used household goods were a staple fixture. Cars and motorcycles, furniture, furnishings, bikes, children's stuff and even pets would routinely feature as columns, with short advertisements of goods listed for sale appearing under each heading.

The decline in local newspapers (along with the rest of the newspaper industry) since the 2000s, accompanied by the rise of digital platforms for monetarized secondhand exchange, has seen the sale of surplus household goods move largely online. Although there are differences between platforms, notably in terms of fee payments and the nature of the market (fixed price or auction bids), the underlying form of the transaction remains much as in analogue mode. Much the same categories of household consumer goods appear for sale, described and listed in exactly the same ways, as single or related item/s of goods – a table and chairs, a chest of drawers, a ladies' bike – as new, an exercise bike, unused walking boots size eight, and so on. Yet underlying these categories are two very different relationships of surplus consumer goods to money.

One of these relationships is the simple form that we have met already. Typically, these are goods displaced by replacement purchases, made superfluous by house moves, or outgrown and/or barely used. Rather than being goods that can be done without for a while, they are possessions no longer needed or wanted. Rather than moving in and out of the category of surplus goods (like coats and Sunday suits), once they become seen as surplus, they are on their way to being divested. The only question is how that occurs and through which conduits they are offered for sale. Invariably, though, they are turned to money through a transaction that passes the property rights in whatever is exchanged to the purchaser.

The second relationship sees consumer goods turn to stocks of money. What appear as simple listings of surplus goods are more accurately seen as surplus goods turned to stocks. This transition has occurred gradually in the UK, but its roots can be traced back to the 1990s and to the advent of the car boot sale.[36] These sales are the UK version of the US garage sale, or 'yard sale'[37], but with one key difference: they bring together what can be hundreds or even thousands of sellers in one place. That convergence of sellers provided the conditions that allowed some consumers to see the potential to accumulate additional surplus goods. For here, depending on the location, were sellers prepared to sell (or unknowingly offering) goods at less than their market value, while the transient and mobile nature of these sales meant that goods acquired at one venue could be stored and then offered for resale in another, thereby exploiting geographical differences in the valuation of particular goods.[38]

Car boot sales have one primary limitation for informal used goods traders: trade is restricted by the temporal and geographical constraints of the physical sale. In short, trade can only be undertaken in one place at one time. The advantages of online platforms over this are considerable: they rely on digital transactions, meaning buyers and sellers do not need to be physically co-present for purchases to be effected; they place goods in a marketplace for as long as a vendor might wish; and they are geographically unrestricted, although, in practice, distance does impinge on purchasing behaviours, limiting many buyers to the relatively proximate. Platforms, therefore, have extended the temporal and geographical reach of physical used goods sales. But they have also created the conditions in which consumers, if they wish, can become full-blown used goods traders without any of the costs of the formal used goods trade (premises, staff costs, credit charges, trading licenses and tax). Much as has occurred in other arenas where platform-mediated exchange has taken hold (for example, accommodation, transport services), extending the used goods marketplace into digital form has intensified what have always been deregulated markets. In so doing, and possibly even more significantly, they have created a new cadre of parasitical consumers, for whom overconsumption, in the form of surplus possessions, is a means to petty speculation and accumulation.

Discard as binning – or, how discard becomes 'waste'

As with used goods exchange, the long historical view shows that the regime governing flows of waste materials from UK households has diversified and intensified from the late 20th century. What was until the late 1990s a municipal waste regime organized around a single bin that contained the contents of the weekly household rubbish collection, is now characterized by multiple bins, each intended to capture particular materials from consumer discard: 'dry' recycling (paper, card, HDPE, PET, aluminium), glass, organic waste (food, garden) and 'residual' waste. In addition to the direct collection of waste materials from households, there are designated sites for disposing of household discard – the various 'bring centres' and household waste and recycling centres provided by municipalities to serve particular areas and neighbourhoods. These accept a host of different goods that may be excluded from the household-based collections: textiles, batteries, electrical and electronic goods, leftovers from home improvements such as old bathroom suites and kitchen units, kids' bikes, old furniture, and so on.

The emphasis on material categories is the clue that this is a regime that turns discard back to resource value. Bins for most consumers are the primary point of contact with the waste management industry, and it is tempting to see these bins as about wastes. But, as later chapters will show, following these 'waste' materials shows that they are anything but wastes. Rather, household bins are the means to the collection of a range of residues, or the material remainder, of consumption. These residues are processed and treated by the waste management industry, prior to entering new value regimes. The most important of these align with energy markets and markets for secondary resources – or materials recovered from finished goods, be they derived from consumer goods or not. What appears to be a regime that manages a particular set of household discards, then, is more accurately designated as the preliminary component of, or supplier of feedstock for, a waste-to-resource regime.

A brief history of household bins

The association of an object taking the form of a bin with humans' discard goes back over millennia. The archaeological record includes prehistoric paintings in the Himalaya of basic wheeled boxes, and similarly shaped remains of wheeled wooden boxes have been found in the volcanic ash of Pompeii.[39] The more recent history of waste organization and management in the towns and cities of the UK is a result of the intersection of municipal and civic responsibilities with public health concerns. Bins for the express purpose of the collection of household rubbish became attached to dwelling structures in the early years of the 20th century for these very reasons. But beyond public health and civic responsibility, like all objects, bins are a form of what is known as material culture.[40] They can be used to tell a narrative, in this case about societies and their relation to the material world.

Prior to the roll-out of bins to households, and for much of the 19th century, the only waste material widely collected from the UK's households was the ash residue from coal fires – the primary source of heating at the time. The most common housing stock of the time – back-to-back terraced housing – had outside toilets, or 'privies'. The privies were built against and into a wall that marked the boundary line of each terrace, with the wall abutting onto a back lane that separated one terrace row from the next. Urban waste management at this time, as well as being about ensuring the separation of wastewater from the drinking water supply, was mostly a matter of managing the mountainous quantities of ash generated by coal-fired and heated cities. So, what was also built into that boundary wall was an 'ash pit privy'.[41] There was one for each house, into which was dumped the ash from each household's multiple fireplaces. Ash privies were accessible from the back lanes that separated each terrace row. The ash was collected and utilized, typically in road building. All other household rubbish, if combustible, was burned on the fire. Otherwise, it was dumped and then scavenged – by other humans but also by pets, birds and rodents.

Come the early 1900s, in the face of increasing public health concerns over open dumping, two-handled, cylindrical metal bins with a lid, known as 'the dustbin', were introduced as the standard means to managing household waste materials – still chiefly ash but also, especially in inner city areas, residues from cooking and food preparation. Scavenging rodents, in particular, had by then been identified as a key disease vector. The combination of heavy metal and lids was not only a means to handling potentially still-hot ashes, it was also a means to deter scavenging – metal being more difficult to chew through. This, plus the relatively heavy weight of the bins themselves, especially when filled with ash, meant that dustbins were harder for animals (dogs particularly) to knock over. So, we can see how the advent of the municipal dustbin inserted a radical separation between human discard and animal, bird and rodent life. Rather than living in an ecological relation, human discards were separated off, through the medium of these bins – as waste to be managed by a specialist waste management industry. The bins themselves were emptied weekly, by 'dustmen' (sic) into a dustcart, initially manually and then, as health and safety regulations started to take effect, mechanically via lifting mechanisms installed on the side of the collection vehicles.[42] This system remained in place across England for much of the 20th century.

The latter part of the 20th century saw the dustbin-dustcart system of waste management in the UK overhauled, through the advent of the wheelie bin. This form of bin was developed in Germany in the 1970s, from a design first used in a Slough-based factory in the late 1960s. In a switch that reflected the move to the plastic age, the material composition of the bin changed as metal was substituted with a lighter, albeit heavy-duty, polymer. More significant is that the size of the bin increased considerably. While smaller (120 L) bins are available, the standard issue wheelie bin is 240 L – well over double the capacity of the metal bin. So, as wheelie bins have been issued to households for more and more material streams (firstly residual waste, then 'dry' recyclables, and then garden waste), the volumetric capacity allocated has been increased, so much so that in many neighbourhoods it is at 720 L/household per fortnight. That alone gives an indication of just how much discard is being generated by UK households. It also highlights the importance of efficient collection systems, for large volumes put a premium on efficient collection. The wheelie bin then is an integral component in a system that the waste management industry has orchestrated to order collection, transportation and disposal as efficiently as possible. The number of 240 L volumetric loads in a municipal area provides the baseline for estimating the collection requirements (in terms of vehicle and crew needs) to service particular neighbourhoods, and for generating the most efficient route algorithms for a given number of collection vehicles in the fleet. They also provide the basis for estimating the yield that can be harvested from these areas. Obviously, not every household will fill its bins for every collection, but many do. Correspondingly, there are environmental campaigners who see these bins as 'waste guzzling' containers, which are as much to blame for contemporary society's waste generation problems as a means to waste's efficient management. Give people a bin – so the argument goes – then they will fill it up and throw things away. But is it really this simple?

The generosity of the bin

Later chapters turn to examine the waste-to-resource regime in much greater depth. For the purposes of this chapter, however, what matters is that the bins that act as the collection points for this regime are one of many conduits through which consumers manage their discard. Unlike the conduits discussed so far, however, they require nothing more from a household than putting the appropriate materials in the relevant bins and then putting them out for the appropriate collection. Undoubtedly, though, these bins – and particularly the residual waste bin, the contents of which remain less regulated than others – provide a convenient means to getting rid of many unwanted things.

As ever when we look at consumption, where there is convenience so there are scheduling pressures. This means that residual waste bins are not just the conduit that households use to manage routine discard – stuff like disposable nappies and pet wastes. They are also the means to dealing with all manner of other things that enter the category of the surplus, often in unpredictable ways. Here, what matters is the ready availability of the bin, or its 'to-handed-ness'. To-handed-ness alleviates the need to make time to do something else with suddenly surplus stuff. So, if I think about the other sorts of things that households who participated in one of my research projects confessed to 'lobbing' in the bin, it would be things like broken kettles and toasters – appliances that 'suddenly decided to go on the blink' and got instantly replaced; odd socks, 'knackered' trainers, items of 'distressed' clothing (that is, with holes, rips and tears), worn-out underwear and hosiery (that is, overly used or impossible to wear clothing); bust cameras; and, in a zone that goes back to the abject (Chapter 2), ripped and overly soiled bed sheets, duvets, battered pillows and pet bedding. This is the kind of stuff that people either know or imagine to be beyond reuse or repair. A strong inference, therefore, is that the material state of an item matters, shaping how it is discarded. Overly used things, displaying too much use, are frequently channelled in the direction of the residual waste bin.

But the residual waste bin is not just a convenient means to managing this kind of stuff. I would argue that the asymmetric social relations that shape this bin make it a particularly generous conduit, for here – unlike all the other conduits discussed – the act of getting rid requires no engagement with another human subject. Instead, the bin's generosity lies in its capacity to absorb unseen and remove a certain amount of material (up to 240 L per fortnight), and to do so routinely and predictably, as per standardized collection rotas. There is no doubt that this capacity is utilized by most households, and that sometimes this use is calculative. This is certainly the case for households moving house. A common experience here is for the intensity of 'binning' to increase with proximity to the move date.[43] With less and less time available to sort and shift stuff through other conduits, all of which require considerable amounts of time, the bin becomes a reliable means to getting rid of a guaranteed volume of material on each collection date.[44] So it is that unwanted but perfectly serviceable things end up in the residual waste stream. But more than this, the residual waste bin is also a means to get rid of what can be usefully thought of as the troublesome surplus generated by overconsumption. These are things that in material terms are perfectly fine, but that, for all manner of social and cultural reasons, are not easy to discard through other conduits. One of the best examples I have ever encountered of this comes in the following story, in which – at her wits end with the evidence of overconsumption littered over the playroom floor – one of the mothers participating in one of my research projects confided that she had picked up and counted the Barbie dolls strewn around the room by her daughters, and then, in a fit of rage, lobbed all 27 of them in the residual waste bin. Getting rid of 27 Barbie dolls through other conduits is open to all sorts of readings (and judgements). The residual waste bin, however, is no judge. Instead, through the waste management system of which it is a part, this bin quietly and invisibly removes these troublesome things, making them 'out of sight and out of mind'. That is its worth for many consumers.

Cascading goods and intersecting value regimes

These three value regimes, and their associated conduits, are analytically and physically distinctive. This has led to the tendency for them to be analyzed separately in the academic literature, which is unfortunate because it obscures that they also intersect, in two ways. One is socially, through practices of valuation. Hints of this have appeared already in this chapter, principally in the ways in which the material qualities of goods relate to the conduits into which items are released. But practices of valuation go beyond this, for people practice divestment in ways that draw on a greater, or lesser, number of conduits. Material qualities figure in this, for sure, but as knowledges of, and commitments to, extending the social lives of things increase, then so does the array of conduits typically deployed. To generalize, the typical divestment pattern of a household with a family living in the UK would involve a combination of the hand-me-down-around economy, charity shops and/or bag drops, eBay, Gumtree and/or Freecycle, Facebook groups, the household waste/recycling centre, bring centres (such as textile and shoe banks) and the residual waste bin to manage their discard. This pattern – often labelled 'easy recycling' by committed practitioners – is added to and elaborated upon by them. Their patterns would typically include a much more strongly differentiated use of specialist secondhand outlets, suited to maximizing the reuse potential in any one item. At the other extreme, there are households – typically ones with limited social ties and networks – who rely exclusively on the bin for divesting things.[45]

The second line of intersection between these regimes is drawn by the passage of items through them. The way this works is that when we follow surplus goods, we see that they funnel inexorably towards the waste-to-resource regime. This is a 'catch-all' regime; bins and 'the tip' (or household waste and recycling centres) are the backstop conduit for all surplus goods but, more than this, the waste-to-resource regime is increasingly a magnet for surplus goods. This we can see when we look at how goods cascade through discard's value regimes.

For all consumers seeking to divest themselves of surplus goods, no matter what these are, there is one fundamental choice that underpins all divestment: either to seek to pass on that good (or goods) to a particular someone else themselves, or to hand the good/s (and what happens to it or them) over to an intermediary. In the first case, that passage could be mediated by money – or not – or by a platform – or not – but in all cases, the good remains in the possession of the original owner until such time as a new owner emerges. In the second case, the good/s are handed straight to the intermediary, typically a charity or social enterprise but, as used goods trading goes increasingly mainstream, it might also be a retailer. Both routes show how goods cascade through discard's three value regimes.

If we start with what we might term 'DIY divestment', this requires consumers not only to hang on to goods deemed to be surplus, but to continue to accommodate them in their homes and, frequently, to continue to put work into their divestment. Very quickly, this can start to become hard work – and work of questionable worth, given that the stuff requiring the work has already passaged to that key category of the surplus. An example illustrates the issues many face, and what then happens to surplus goods.

A couple of summers ago I was at a neighbourhood barbecue, where I was asked by an older female acquaintance whether I would like a set of cane conservatory chairs with soft furnishing style cushions. Apparently, the set had been up on Gumtree and then Freecycle for over a month but had attracted no interest, so the couple were anxious to shift it, especially since their newly ordered furniture would be arriving soon. Although I wasn't particularly interested in said furniture – not having a conservatory – I was sufficiently intrigued by its potential possibilities to go along to have a look at it. Described to me as 'ten years old and a bit bashed', it turned out to be exactly that. Evidently, I was its 'last chance saloon', for when I declared no interest in taking these items, the couple then decreed that it was time for the stuff to be loaded in the car and taken to the local recycling centre. In such a way, we see how the failure of things to be revalued through multiple conduits (Gumtree, Freecycle and social networks – tellingly in the order money, followed by free exchanges, with the hand-me-down economy as a last resort) results in their cascading to a conduit that is an entry point for the waste-to-resource regime. We also see that there is a relatively tight window in which revaluation has to occur. Time and space begin to compress here, rendering it imperative that these surplus goods are moved on, to be handled by another conduit. As compression increases, so the convenience offered by the conduit associated with the waste-to-resource regime becomes increasingly attractive. This happens time and again with surplus goods. Unless these goods are regarded as stocks by consumers, there is a limited period in which they will be held by those seeking to divest themselves of them. Failing to find a recipient results, inexorably, in the passage of surplus items to the conduits of the waste-to-resource regime, be these bins or local recycling centres.

Cascading goods are not just a characteristic of DIY divestment. Similar patterns are to be found with the intermediary route. Take donations to charity shops as an example. What happens to these is that they are first sorted in back rooms by teams of volunteers. Sorting here is by category and quality, so, women's, children's and men's clothing, books, jigsaws, toys, CDs, DVDs, and so on, then categories within those, if appropriate (so, jackets, trousers, coats, shirts, blouses, skirts, dresses, and so on); then by condition and by brand if appropriate. Much as with consumer valuations, volunteer valuations are variable – so, what is deemed to be of good quality, or recognized to be of value, varies according to who is doing the valuation work. But in general, the aim is that only good-quality stuff goes out on the shop floor (or is offered on eBay), because it is this that will realize the most money. The rest is divided up. A small amount of stuff deemed to be of reasonable quality might be directed to another store in the charity's local network, particularly if it is stuff that is known to sell well there. More commonly, any clothing that does not pass the shop floor quality threshold is bagged up and redirected to textile recyclers. In the charity shop that I volunteered in, mountains of donated clothing actually ended up not on the shop floor but rather at the charity's giant clothing sorting plant (the black hole of Wastesaver of the Preface), where it was further segregated and graded, prior to being sold into the global reuse and recycling markets for clothing and textiles.[46] For the much smaller amount of goods that went out onto the shop floor, there was a similar window of revaluation to that I described in relation to the conservatory furniture. In the instance of the shop I volunteered in: three weeks. If the good had not found a buyer in that time, it was removed from the shop floor and then sent either to another branch in the network or directly to the warehouse facilities, thence to cascade to the waste-to-resource regime.

The hard reality of much used goods exchange, then, is that while this might look like saving things from becoming waste by extending the social life of things, when we follow things through these conduits it becomes clear that most things only have a finite time for that potential to be realized. As a result, many used goods fail to realize their potential for revaluation, rapidly becoming feedstock for the waste-to-resource regime. This is a function, on the one hand, of used goods trading acting as a buffer zone, mediating between consumers' acquisition and displacement of goods and the capacities of the waste-to-resource regime to endlessly absorb consumption's remainder. Given capacity constraints, there are limits to what can be held in that buffer zone at any one time. On the other hand, it is also an effect of needing the right buyer to come along at the right time. In circumstances defined by the unpredictability of the availability of particular goods, that can be a challenge, for consumers and traders. Ultimately, then, holding on to goods in the hopes of finding a buyer gets trumped by the sheer weight of goods being cast out by consumers, and by the capacity of the waste-to-resource industry to turn this weight into money. In the end, time runs out for all things, and discard realizes value through its recovery value.

A final note – waste prevention or parasitical consumption?

I want to end this chapter by reflecting on the implications of the current enthusiasm for used goods exchange as a form of waste prevention behaviour.

The argument that frequently gets rehearsed here is that sites of used goods exchange are a means to the valorizing of reuse, be that intentional or an indirect benefit. Seen as distinctive sites, they are that, of course. And reuse sits at the pinnacle of the waste hierarchy, as indicative of waste prevention behaviours. But this is to neglect the dynamic of consumption and the waste-to-resource regime that I have articulated in this chapter, and the ecological relation that sits at the heart of that connection. Thought of in ecological terms, used goods exchange is not an alternative form of consumption but rather a parasitical form of consumption. Like all parasites, it depends on a host for its existence – in this case, on divestment, or on goods being released by consumers. Without the endless replenishment of new goods, then, there can be no renewal of the supply of used goods. Rather, and as per the parasite, used goods exchange depends on new cycles of commodity production and consumption.

This matters not only because it is important to see used goods exchange for what it really is, but also because something quite dramatic is beginning to happen to reshape the dynamic of this particular host–parasite relationship. Until recently, I would argue, the relationship between used goods exchange and its host in the UK has been relatively in balance. That is, discard's three regimes of value have acted to absorb consumer discard without driving increased levels of demand for new goods. Now, in the name of waste prevention behaviours, and in an attempt to push what are seen to be more sustainable practices on retailers and producers, the demand for used goods seems set to increase, threatening the balance of that host–parasite relationship. In the drive to create a circular economy, used goods exchange is being 'mainstreamed'. It is early days in this, but already we can anticipate how this is likely to play out. If we take the case of Ikea as an example, in a clear endeavour to drive more acquisition, consumers returning certain furniture and storage goods will be given vouchers, aligned to the reuse value (determined by condition) of the goods that are returned. In what amounts to a consumer reward scheme, exchanging used goods for vouchers seems set to drive more consumption (of used or new goods), for what point is there in a voucher unless it is exchanged for goods, or sold on to someone else wishing to make a purchase? Potentially, too, it will speed up the return-acquisition cycle, for it will not be lost on consumers that better-condition goods command a higher return. Rather than slowing consumption down, a clear possibility is that vouchers for used goods will become the means for consumers to engage in ever-intensifying, fashion-driven cycles of return and acquisition.

Step beyond the case example and mainstreaming used goods exchange will increase the competition in the used goods market. Demand will increase, for with bigger and bigger retail players entering the market the competition to capture both quality and volume will intensify, not least as retailers will be required to evidence their waste prevention activity. In the face of this, social economy actors and consumers seeking to act as traders will find quality stocks increasingly hard to come by, while consumers doing DIY divestment for free on certain platforms may find themselves faced with the counter attractions of retailers offering them rewards for the surplus goods they might currently be offering for free. So the entry of big retail players to the market will be disruptive, but only to the other conduits in the market of seeking to revalorize things as things.

We might speculate, then, that when used goods go mainstream, practices of valuation will ensure that only a limited number of (very good condition) things will actually get offered for exchange. Much more common will be that things are counted as accepted for reuse exchange, but their condition will foreclose their being offered on a sales floor or a platform. Inevitably, this will mean that they will cascade down to the waste-to-resource regime. In such a way, what looks to be, and will count as, waste prevention, is more likely to be a means to intensifying the connection of consumption to resource recovery. Further, what looks to be a future characterized by reuse exchange is, I would suggest, more accurately seen as a means for retail capital to use used goods to drive more consumption, while simultaneously being the means to the waste management industry capturing the recovery value latent in these goods. This is 'closing loops' mainstream style. Rather than consumption being an open pipeline for a whole range of actors beyond the mainstream economy to capture a slice of the value in used goods, in the name of circularity that pipeline is being shut down by capital and enclosed within the mainstream economy.

5

Recommodifying Discard

Or, the challenges of turning discard into an economic good

Since the beginnings of capitalism, discard has been recommodified. There is nothing new about this. Scratching around in municipal dumps, searching amid the mountains of abandoned detritus for materials of value – stuff that can be salvaged, accumulated and sold – is an activity that has been found the world over for centuries. It is a tactic for 'getting by'; an age-old indicator of livelihoods marked and defined by living in conditions of extreme poverty. Scavenging provides the defining contemporary image of the landfills and dumps of the global South: one of women and children sifting carefully, methodically through mountains of rubbish while living in its immediate proximity, surrounded by societies' detritus. Harrowing as such images are to western eyes, it is important to recognize that scavenging and salvage activity was as much a feature of the dumps that littered the landscape of 19th-century England as it is of today's cities of the South.[1] One only has to turn to some of the novels of Charles Dickens to find precisely the same close association of extreme poverty and rubbish-dependent livelihoods.[2] The more unequal societies, the more significant discard is for livelihood strategies, with large numbers dependent on rubbish livelihoods.

The recommodification of discard, however, is not just about scavenging and salvage, notwithstanding that this is the focus for most of the academic literature in the social sciences. It also encompasses the trade in used goods and the transformation of discard by varying combinations of labour and capital, such that discarded things, or their constituent components and materials, are opened up to be bought and sold. In such a way, scavenging and salvage activities connect to the intricacies of the value chain in scrap.[3] Typically, all this economic activity is now lumped together under the catch-all umbrella term of 'recycling', but that term includes a range of activities that, in actuality, are distinct. Not only do they operate in distinctive markets, they also have widely differing implications for debates about sustainability. At one end of the range, and without doubt the most sustainable of these activities, is reuse, where discarded used goods are resold to fulfil the same function that they were originally designed for. Discarded used goods, however, are also sold into repair and recovery markets; so they are sold to be taken apart, or cannibalized. Sometimes – as with e-waste, for example, or cars – they can be used to source spare components, but in addition, they are dismantled to recover their constituent materials, insofar as this is possible, for many manufactured things are made through processes that combine and engineer materials in ways that are difficult to reverse (Chapter 3). This last activity is the essential precursor to recycling. So that materials can be recycled in further rounds of manufacturing, they first have to be recovered from already manufactured things. Only then can they be sold into the commodity markets. Recycling, then, depends on the prior destruction, or disassembling, of things and their separation into clear streams of sufficiently distinctive, recovered materials; further, it only occurs once recovered material has been incorporated into manufacturing processes.

Used goods trading and recycling activities have been around since the birth of capitalism. But, unlike scavenging – which remains a localized activity, fixed in the spaces where people live alongside dumped discard – these activities have changed considerably in recent decades. What is new about their configuration now, aside from the changes in the nature of consumer discard and the materials that are to be found in that discard, is the scale of these activities, the geographical location of recommodification activities and the processes that are involved in the transformation of discard. These are the concerns of this chapter.

Underlying all of this change is globalization and, more particularly, its conjuncture with what is known as 'globalization from below'. The familiar story of globalization for audiences in the global North is one of the liberalization of the economies of the developing world from the 1980s; of the consequent outsourcing by transnational corporations of much assembly and low-value manufacturing activity from the global North to the economies of South-east Asia, and then parts of South Asia; and of the increasing reliance of the economies of the global North on service activities – retail, finance, health, education, and a raft of consumer services such as hospitality and tourism. This is globalization as seen from the perspective of the global North. Less familiar is what is known as 'the view from below'. But, as this chapter shows, this perspective is critical to understanding how discard has been recommodified, and where that happens.

At the heart of globalization from below is entrepreneurial trading. Academic research on the economies of the global South and China has made visible networks of traders – mostly Chinese and African – for whom the structural conditions of globalization have provided the opportunity to develop a highly lucrative North–South, South–South and China–Africa trade in used goods, counterfeit goods and scrap materials.[4] Traders use short-stay visas to build business, focusing their activities in key entrepôts, which are the main conduits through which such trade funnels and within which trading relationships are forged.[5] The broad conjuncture behind all this is comprised of cheap 'back run' shipping costs that result from the primary flows of finished goods to the consumer markets of the global North[6]; the development of 'free ports' in the economies of the South, to enable tariff-free trade; and demand for used consumer goods in the economies of the South, where the structural conditions of globalization mean that the vast majority of people do not have access to the goods that they make for global markets, and either cannot afford to or may not wish to purchase many goods that are produced locally. What middle-class consumers in the global South can afford to buy, however, are the cheap, branded goods discarded by consumers living in the global North, or counterfeit goods. In such a way, used consumer goods flow south, to appear on the street markets of the cities of the South; to be sold as used goods, to function as a reservoir of component parts for the extensive repaired and reconditioned goods sectors of these economies; and to act as a generalized source of recovered, or secondary, resources that in turn are sold into local commodity markets and thence into local manufacturing activity – not export-oriented assembly and manufacturing, in which 'virgin' resources comprise the bulk of the raw materials input. Many scrap materials recovered in the global North (be that from consumer goods or industrial goods) flow in much the same way, functioning, for the most part, as a supply of cheap secondary resources for local manufacturing activity in the global South.

The global recommodification of discard does not get a good press, most particularly from environmental campaigners, but also, to a lesser extent, from development specialists. For environmental campaigning organizations, a subtle elision works to ensure that discard is rendered the equivalent to waste. Here, the global trade in used goods and scrap materials is presented as a global trade in wastes, which dumps the wastes of the global North on the people and environments of, in the main, the global South. No matter the waste – be that a generic category such as industrial or consumer wastes, or a material category such as e-waste or, the current bête noir, plastic – it is the same story – one of the dumping of toxicity.[7] There are elements of truth in this story, of course, for this discard is classified as waste in the global North, and also increasingly in China, and there is much in this discard that undoubtedly is toxic when released uncontrolled into the environment. But what the story downplays through its focus on waste is the entire process of recommodification, and thus the raison d'être for why goods and scrap materials flow in these directions in the first place. It therefore fails to recognize that the waste that appears in the South is mostly the residue of processing used goods and scrap materials[8], conducted in conditions where environmental regulations are laxer than in the global North – a situation that allows for less technically controlled processing and for the dumping of residues. A further consequence is that, in focusing on the origin of these wastes and downplaying the significance of the recommodification of discard, these accounts pave the way for the more dramatic representations that underpin environmental campaigning – slogans such as 'toxic colonialism' or the 'dark side' of globalization.

A brief (and critical) note on toxic colonialism

One of the best illustrations of the toxic colonialism representation in recent years is the Trafigura/Probo Koala case of 2006, in which 528 m^3 of toxic residue from processing coker gasoline was dumped at an open-air dump on the outskirts of Abidjan in Ivory Coast. Framed through the lens of environmental justice, for campaigners and many media outlets in the global North, this was narrated as a straightforward story of toxic dumping in which a multinational could be inferred to have dumped toxicity on impoverished people in the global South, with deleterious consequences for their health. Trafigura, however, is not a multinational in the standard sense of that term (think car manufacturers). Rather, it's a global commodities trading house. Commodity traders buy and sell resources; oil, metals, coal and agricultural grains are major areas of their business. They are the real-economy counterparts to the financial traders whose work came out of the shadows as a result of the global financial crisis, yet until very recently, commodity traders have existed in the shadows of the global economy. As investigative journalism has made clear, though, these traders are the glue to the global economy, enabling resources to flow, including from regimes that are seen to be geopolitically risky, or worse – pariah states. So, behind the Trafigura case is a complex story of resource trading and its intersections with residue trading.[9]

In 2005, Pemex – the Mexican state oil company – offered coker gasoline to the open market at a rock bottom price. The company had run out of storage for coker gasoline and needed to shift it fast. Coker gasoline is a residual cracker product containing high levels of sulphur that need to be removed for it to be sold into other markets – chiefly as a fuel for power stations and cement works. Removing that sulphur is a very costly process when it is done in parts of the world that have strong environmental controls, but there is another process (caustic washing) that has the same result that is way cheaper. Its downside is that it results in a large amount of toxic residue. That process is banned in many parts of the world, but not all. Email correspondence from Trafigura showed that the trading house saw in Pemex's coker gasoline the opportunity to make a considerable amount of money (an estimated $7 million), by using the caustic washing process. However, their speculation started to encounter problems when they failed to find a port that would accept the cargo for processing. Desperate to avoid the oil refinery processing route, which would have wiped out their anticipated profits, they opted to buy a vessel offered in the ship demolition market and use it as an offshore processing site. Enter the Probo Koala. The caustic washing process took place aboard the vessel, giving Trafigura a product that it could then sell but also leaving it with a residue (held in the slop tanks) that was classified as a toxic waste. Having been quoted an alleged price of $700,000 by a toxic waste management company in Amsterdam to take that slop cargo and dispose of it safely, Trafigura sold the cargo to a company based in Ivory Coast for $20,000. It is that company that dumped the residue at the open-air dump outside Abidjan. In yet a further twist to the tale, the Probo Koala (itself having been renamed) was sold to ship scrapping firms in first India and then Bangladesh, but its recent history contaminated it by association and in both cases the ship's beaching was blocked at the highest level. It was eventually scrapped in 2013, in China.

★ ★ ★

Slogans such as 'toxic colonialism', in my view, are unhelpful. They skate over the complexities of resource and residue trading that are central to understanding the recommodification of discard, and they elide discard and waste. They also occlude that what is going on in this trade is a global reconfiguration of supply and demand, in which it is the conjuncture of cheap discard in the global North with globalization from below that has been critical. It is this conjuncture that brought about the harvesting of the discard of the global North by traders from the South and China, to become goods and resources for and in the South and China.[10] There is nothing particularly colonial about this arrangement. Rather, if there is a slogan that captures the situation in the global South more effectively, it is 'reverse imperialism', whereby postcolonial economies – many of which bear the scars of a resource-centred development designed to support the economies of the 19th-century imperial powers – now find themselves facing the challenges of an economic development reliant in considerable part on secondary resources – resources that emanate from those very same former imperial powers.

The challenges of recommodifying discard for the economies of the South have mostly been considered by development specialists, who, unlike environmentalists, recognize discard for what it really is: discard, not waste. Their focus has largely been on the used goods element of recommodification and its relation to both the conditions and the possibilities for local domestic industry and manufacture.[11] Debate here is clouded by the realities of economic liberalization for many of these economies.

This is widely recognized to have led to the development of economies that are highly dependent on swathes of relatively low-value assembly and manufacture activity, all oriented to export markets. In short, economic liberalization turned these economies into 'factories for the world', where what that really meant was producing cheap goods for the consumers of the global North. The challenge of economic development for these countries, then, is to move their economies up the value chain. But that remains difficult for many, especially given that the main value-adding activities of the major transnational corporations (branding, marketing, research and development) continue to take place in the global North. These challenges are recognized to be compounded further by imports of large volumes of cheap used consumer goods. Given the reach of global brands, these goods are often more popular with local consumers than locally produced goods, the markets for which therefore contract, with inevitable consequences for domestic manufacturing activity. It is these circumstances that have led many developing countries to ban the importation of used consumer goods. Yet, such bans continue to be easily circumvented through indirect trade and illicit trade and by routing goods through proximate free ports. Rather different is the trade in scrap materials, for scrap has long been recognized either as an essential component of basic manufacturing (as, for example, with electric arc steel manufacture) or as a means to accessing resources cheaply, and thus as the cheap route to the development of domestic industry. This has been particularly important for countries whose currencies are not tied to the main international trading currencies and/or who lack access to key resources. The best, indeed major, illustration of scrap-reliant economic development is China.

A brief note on China

The emergence of China as a major economic force (Chapter 3) fuelled huge demand for resources and instigated a commodities prices 'supercycle' that lasted through the entire boom period of Chinese economic growth. Chinese demand could not be met by primary resources alone. So, China became the main global importer of the major categories of scrap material during the high period of globalization from the end of the 1980s to 2010.[12] China's demand for scrap can be traced back to the late 1980s, when Taiwanese scrap metal and paper traders set up the first recycling companies in China, and it led to major concentrations, or agglomerations, of recycling activity in Guangdong, Jiangshi, Zhejiang and Shangdong provinces.[13] As recently as 2010, imported paper scrap accounted for 50 per cent of the paper used in China's paper mills.[14] From 2013 onwards, however, the Chinese state introduced a set of regulations that cracked down progressively harder on scrap imports and that attempted to reduce Chinese manufacturers' dependence on recovered materials. These measures culminated in 2018 with a complete ban on the importation of 24 grades of solid waste, including unsorted mixed paper and post-consumer waste, and with the imposition of a contamination threshold on import loads that was impossible for consigners to meet. The result was the most disruptive event in the global trade in discard in 20 years.[15] It led to vast quantities of post-consumer waste in the global North being without a market and being landfilled and incinerated, and to the displacement of plastics processing capacity elsewhere in the world – to Malaysia, Thailand and Indonesia – as well as to increased volumes of material heading to other major destinations, notably Turkey.

In part, the ban sought to upgrade China's economy (Chapter 3) and to distance China politically from its image as processor of the garbage of the global North.[16] It also reflected lobbying by China's domestic recycling industry and the strong growth in Chinese consumption that was generating its own post-consumer waste. Yet what has transpired recently, alongside political reiterations of the ban, is the gradual relaxation of that ban in relation to certain recovered materials, notably key metals (brass, aluminium and copper), ferrous and stainless steel scrap, and also paper.[17] Regarded by commentators as a classic example of Chinese 'face saving'[18], this partial about-turn (which continues to require that imports meet high purity thresholds) is a means of recognizing China's manufacturers' continued dependence on recovered materials, but also of externalizing the costs of producing those recovered materials elsewhere. China, then, has joined the discard displacement party that has characterized the global North under globalization, but it is doing so from an already strong base in utilizing recovered materials in domestic manufacturing.

★ ★ ★

The remainder of the chapter charts the global shift in the recommodification of discard through a focus first on what is the most familiar of goods, as well as that which has attracted the most interest from academic researchers: textiles or used clothing.[19] It then turns to the work that goes on inside a textile sorting factory, highlighting how value is created from mountains of clothing discard. A look inside the world of textile sorting allows for opening up the labour process in the recommodification of discard. For very good reasons, the world over, the sorting of used clothing discard continues to rely on large amounts of human labour, while the further processing and treatment of this discard has barely changed in 200 years. As such, used clothing provides an example of the tendency towards inertia in the recycling/recovery sector, while simultaneously flagging up its converse: the challenges of innovation. These challenges provide the focal point for the rest of the chapter. Undeniably, the major innovation in the recovery/recycling sector in recent decades has been the advent of capital-intensive, but largely mechanical, processing technologies, for the most part located in the global North. This innovation developed primarily in response to changing environmental regulation, which classified discard as waste and sought to stop waste from becoming pollution by sequestering waste closer to home. I examine these innovations and highlight their connection to, and often dependence on, global export markets. I then turn to consider the potential of a new round of discard-based innovation in the global North, based on chemical and biological transformations. Mechanical, chemical and biological transformations all highlight the key challenges that face the recommodification of discard through recycling. Regardless of how that is done, these are: technological scale-up, achieving standardization of output, and commercialization in the face of market competition with established products produced from virgin, or primary, resources. The challenges are compounded by the inevitably highly dispersed nature of much discard, which produces resource geographies quite unlike primary resource mining, making the costs of harvesting discard considerable.[20] I conclude the chapter by sounding some notes of caution regarding the potential for such innovation to recast economies in circular ways.

Recycling textiles in historical and contemporary perspective

From Batley to Panipat: the global shift in textile recycling

Textiles play a central role in accounts of the development of industrial capitalism, and rightly so, for what happened in the cotton and woollen towns of Yorkshire and Lancashire in Northern England in the late 18th century led to the demise of handloom weaving and to the emergence of what became known as the factory system (in the form of mills) and agglomerations – or geographical clusters – of related economic activities, in this case spinning fibres and yarns, weaving cloth and the allied manufacture of a whole host of finished textile goods for industrial as well as consumer markets.[21] Less commonly told is the story that lags this slightly, of the development of a textile recycling industry and its products: shoddy and mungo.[22] This story is not just important as the foundation story for industrial-scale recycling. In its details it contains important parallels with, and holds important lessons for, current regulatory interventions that, in a quest for enhanced resource security, seek to intensify the recommodification of discard through recycling.

Industrial-scale textile recycling emerged in the early years of the 19th century in what has become known as the 'Woolopolis' of West Yorkshire – primarily Batley and Dewsbury, but also Leeds, Bradford and Wakefield.[23] As with many of recycling's innovation stories, the much recited narrative that surrounds this particular story has a hero at its centre, in this case Benjamin

Law, who, on travelling to London to source wool, came across a saddler stuffing a saddle with torn woollen rags, scarves and sweaters. So the story goes, Law saw a future for this material: he recognized the potential to use it to produce a 'renaissance yarn', which could offer a cheap substitute for wool. The story, of course, has more than a whiff of tales told with the benefit of hindsight. In its form it is a classic retrospective heroic narrative, where what transpires allows for a backwards retelling centred on an important visionary figure. So, what often gets left out of a story that also includes technological innovation – in the form of the reconfiguration of existing machinery such that rags could be shredded – is just why the quest to produce a 'renaissance yarn' was so compelling. Aside from the obvious motivation for a capitalist of higher rates of profit on cheaper raw materials, the other key imperative was shortages of raw wool imports, brought about by rising demand for wool products. In other words, it was the scarcity of raw materials that provided the key economic conditions of possibility for the birth of industrial-scale textile recycling in this part of Yorkshire. For the mills to keep up with demand, they needed to find a way to use resources more efficiently.

Subsequently, as Law's innovations took off, this part of Yorkshire became the centre for textile recycling in 19th-century Europe. It was supplied by an unknown number of 'rag and bone' collectors, sourcing used garments from the populations of the major English cities as well as Europe. By 1873, there were 50–60 mills in Batley alone, with 3,000 looms producing shoddy yarn. Yet, multiple questions hung over the industry, notably with regard to the quality of shoddy as a product[24] and the labour conditions of its production, where the term 'devil's dust' was coined to indicate the lung disease that went hand in hand with working with ragging machinery.[25] A contemporary account indicates the ambivalence around shoddy and the town at its centre, with Batley described in an influential periodical of the time, *The Westminster Review*, as 'the famous capital ... whither every beggar in Europe sends his cast-off clothes to be made into sham broadcloth for cheap gentility'.[26] Similarly, Marx and Engels identified shoddy with poor quality, cheap products that became threadbare in less than a fortnight. While such cloth was undoubtedly produced by unscrupulous, cost-cutting manufacturers, academic research has shown that others devoted considerable time to experimenting with combining shoddy with 'virgin' wool and to establishing the appropriate combinations for different cloths, without losing business.[27] Indeed, it is argued that the widespread use of shoddy and skills in its blending were critical to Britain's growing share of global wool textile exports in the decades between 1880 and 1913. That expertise was recognized by competitors, with it being stated that, 'in some parts of Europe, especially Yorkshire, manufacturers are enabled to get more poor material into yarn than in any other part of the world, for the simple reason that they are highly skilled in the art of combining'.[28] Suits, military uniforms, carpet lining and stuffing products for furniture and furnishings were among the major product lines to incorporate shoddy, with goods exported widely to Europe and the US.

In terms of lessons for today, then, this part of West Yorkshire was the first to experiment in what industrial-scale resource efficiency meant for product manufacture: the skill of blending recovered and raw material fibres in sufficient proportions such that the material combination did not result in a perceptible diminution in product quality.

The 'long' 19th century was the high point for West Yorkshire's textile industry, and for shoddy manufacture within that. Although shoddy manufacture persists, it is a vastly reduced activity, with a small number of – mostly Pakistani and Indian-owned – mills producing stuffing for use in carpeting, mattresses, auto panelling and speaker systems.[29] Batley's contemporary counterpart is instead Panipat, 90 km north of Delhi, which, until recently, has been the undisputed textile recycling capital of the world. The parallels between the rise of the industry here and in Batley are considerable. Like Batley was, Panipat is at the centre of the Indian woollen industry, producing a range of home furnishing products, including durries, woollen carpets and blankets. In a touch of irony, it owes that economic focus to the British East India Company, who shipped raw wool from Southern India to Panipat to produce blankets for the British Army.[30] As with Batley, the turn to using recovered textile yarn in Panipat was precipitated by increased costs in virgin wool imports in the 1990s. But rather than invent machinery, and in a move that shows that the trade in secondhand goods applies as much to capital goods as to consumer goods, Panipat's manufacturers bought up machinery from the then declining Italian woollen cluster centred on Prato (that the Italians had in turn bought from England) and moved it to India.[31] Hence, in the 2010s, antiquated machinery that would more than likely have been recognized by Benjamin Law was still to be found working in Panipat's mills.[32] By the early 2010s, some 400 mills were processing used clothing imported from Europe and the US to produce shoddy yarn.[33] At its peak, Panipat's shoddy mills were an INR 700–1,000 crore (>$100 million) industry, accounting for 90 per cent of the market in global relief blankets.[34] Like Batley, however, the same question marks around quality persist with these shoddy aid blankets, which, in echoes of 19th-century commentary, have been described by contemporary researchers as composed of 'a semi-felted woven mesh of short fibres, dirt and dust barely stuck together, soaked in a chemical cocktail … of unprocessed waste oil, possibly bulked out with salt or starch' and as falling apart in a few weeks.[35]

The highpoint of shoddy manufacture in Panipat was between 2008 and 2010. By 2017/8, business journalists were signalling a clear decline, with only approximately 100 mills still operating, producing 300 tonnes of yarn per day (a 25 per cent reduction on 2008 figures).[36] The precipitate for the decline is widely seen as competition from China. Fleece produced from virgin polymers is seen to offer not only a superior quality product (softer, lighter, available in different colours), but is also cheaper than shoddy (INR 80–250 cf. 90–300).[37] In response, the Panipat mills that remain in the relief blankets market are reported to be shifting to virgin polyester. The fortunes of Panipat's shoddy manufacture, therefore, serve to highlight the market vulnerability of goods containing a high degree of recycled content to developments in primary manufacture. They are also suggestive of a tendency to inertia compounded by a reluctance to innovate and reinvest.

More broadly, the decline in shoddy manufacture in Panipat is seen to spell big trouble for the growing global volume of discarded used clothing (Chapter 3), for which it has been a key global hub. Business journalists have rightly emphasized that, even at its peak, Panipat's mills did not have the processing capacity to absorb the growing tide of used clothing discard – and that was before the rapid expansion in Chinese used clothing exports.[38] They posit a perfect storm of rising volumes of discard but contracting markets to absorb and process them. Equally threatened by the decline of Panipat are the sorting factories for used clothing in the global North, and their allied sourcing networks. The next section takes a detailed look at what goes on inside these sorting factories, both to unpick textile recovery in more depth and to use that to establish some general principles about resource recovery, which hold regardless of the material being recovered.

Inside the sorting factory: what happens to used clothing

The supply chain for Batley's shoddy mills was highly dispersed, comprising numerous 'rag and bone' collectors working in streets and neighbourhoods in England and beyond and a dense network of small rag-sorting workshops. The counterpart to those sorting workshops is still to be found in Panipat[39], but in today's global textile reuse and recycling industry, the first link in the value chain is the textile sorting factories of the global North. These – often giant – plants show that textile recovery activities have been thoroughly incorporated into industrial-scale capitalism. Furthermore, the costs of that supply chain have been driven right down. Rather than being supplied by dense networks of small-scale traders and workshops, their supply chains are serviced by largely unwaged labour, in the form of volunteers and consumers, for the garments and cloth that end up at the sorting factory gates come in the main from charity shop discards sorted by volunteers and from textile 'bank' collections and return schemes, located in places like supermarket car parks, retail stores and household waste recycling centres, where consumers drop off their discards for free.[40]

In that discarded clothing and textiles are moved through these factories on conveyor belts, textile sorting factories are mechanized. But mechanization in textile recovery is only partial. Much like any factory based on lines of conveyor belts, the speed of the belt is critical to the efficiency of workers and the productivity of the sorting factories[41]; and here, it is a veritable army of workers – mostly women[42] – who are central. Their task is to stand by the belts, quickly assessing garments as they pass, pulling them off the belt and placing them in bins. The bins correspond to categories, chiefly of particular types of garments and materials. Rather than an assembly line, then, this is a sorting and segregation line, which works to constitute classificatory order out of the mountains of undifferentiated used clothing that arrives at the factory gates. How the sorting system works goes something like this.[43]

Inside a factory in London there are six conveyor belts. The first two are used to pull off the best and the worst clothing from the incoming material, with 'pickers' assigned to pull off either category. The 'best' garments are then directed to an area of hand sorting, where each garment is assessed individually. This is instructive – it tells us that real quality cannot be assessed at the speed of the belt. This type of used clothing is known in the trade as 'crème'. It is destined for the reuse markets of Eastern Europe, where Cream is a retail chain store selling good-quality, mass market British brands such as Marks & Spencer and Next at rock bottom prices.[44] To get sent in this direction, however, the clothing has to pass the quality control thresholds of those working in this area, and to do that it must be assessed as either Grade 1 (unworn) or Grade 2 (very good condition). Anything that fails that categorization goes back on the belts. In contrast, the worst garments – the type of things that are covered in blood, vomit and excrement, or mud, or which are soaking wet – are destined at best for the rag industry, to be shredded to produce industrial-grade wipes, or at worst are classified as 'waste' and directed to landfill. Working on the initial belts, then, pulling off the worst grade of clothing, is dirty and potentially dangerous work. It is so not only because of the state these used garments are in but because in this factory, gloves are not part of the job. This is because assessing garments swiftly relies not just on sight and smell but also on evaluations of material type, where unmediated touch is the fastest means to categorical assignment.

Beyond the first two belts, the remaining four are used to pull off garments destined for the core global secondary reuse and recycling markets, namely West Africa and Asia, respectively.[45] The precise number of grades that are used to order this sorting is closely guarded, but firms will admit to at least 160–200. Beyond the basic distinctions of geographical destination and likely end use (reuse or recycling), the first-order categories are used to differentiate garments by type (blouses, shirts, trousers, men's, women's, children's, and so on) and by fibre type (for example, wool, cotton, linen, silk, polyester), but also feel. In this way, categories like 'silky' come into the mix. These do not translate straightforwardly into a material type. With no time to search for and read the label, garments need to be assessed at speed. So, this is skilled work but also subjective work, for one person's 'silky' is not another's. The grades that lie sorted in the bins, in turn, are consolidated and translate into goods. Goods here are tightly compressed 45 kg bales that are the basic unit of trade in the export market for used clothing, and the unit that choreographs the industry's products with the workhorse of global trade: the standard 20' shipping container. Bales are consolidated into orders (typically corresponding to one or more container/s), which are graded, via a price list, for buyers, who are the key intermediaries to wholesale and retail markets in Eastern Europe and the global South.

As with any retail market, buyers in textile reuse markets not only operate in distinct markets but also occupy different niches, commanding different order books and working to different wholesale and retail price points. Given currencies and buying power, buyers operating in the markets of the global South are not buying at the same price point as buyers from Eastern Europe. They buy the lower-graded quality bales in the main. In Eastern Europe, Cream's accounting for but a fraction of the crème grade also means that much of what is destined for Eastern Europe is bought to be resorted, inside yet more sorting factories.[46] Some of this clothing does end up in other secondhand clothes stores, but only the best. So, large amounts are re-exported into the global market, meaning that Eastern Europe is not only an end market for used clothing but a key transit hub.

Regardless of the market they are operating in, though, the problem that all used clothing buyers face is that they are buying, by definition, non-standardized goods. Rather than purchasing x amount of y clothing good (in a range of sizes and colours), as a fashion buyer for a clothing retailer in the global North does, they are buying graded bales of a particular type of garment – trousers, blouses, t-shirts, bras, for example. Each garment in the bale then will be a different style/make/size, and multiple brands will be jumbled up together.[47] As such, buyers cannot control the supply chain as they do with clothing manufactured for sale in the mass markets of the global North. There they determine not only order books but also production runs and schedules in the garment factories in the global South. What has evolved, therefore, in terms of buyer–supplier relations, is more of a co-dependent trust relation between factories and their key buyers.[48]

Buying blind and at a distance, each buyer must assess the commercial risk on their order. Then, orders translate to shipping containers. The biggest players in the market pre-order multiple container shipments spread multiple times across any year, thereby guaranteeing to absorb large volumes of a sorting factory's output. The sorting factories, therefore, depend on satisfying these buyers, so there are strong pressures on the factories to keep quality levels acceptable. Smaller traders, however, and those starting out – of which there are many given the entrepreneurial trading that prevails in the global South – either order 'less than a container load' or are dependent on sourcing bales from importers, effectively 'middlemen'. Some of these, at least, will comprise cheaper orders of lower-quality goods and/or they may include multiple bales of potentially harder to sell garment types alongside what is known in the trade as 'nuggets' – or stuff that will sell like the proverbial 'hot cakes'. Since it is impossible to know what exactly is in the bale until the bale is split open, there is more risk at this end of the market. Research conducted with Nigerian traders[49] gives an example of the level of commercial risk involved. It shows how the losses on an order that ended up comprising too many bales of pairs of men's trousers were only partially offset by bales of t-shirts that sold well. It also highlights why, in conditions of asymmetric knowledge, traders have turned to sending apprentices to work inside textile sorting factories, particularly in the UK.[50] Not only do the apprentices have a better knowledge of the finer distinctions in the end markets of street fashion, which can be used to improve the quality and consistency of a factory's garment classifications, they are also there to develop trust relations with the sorting factories and to grow buyers' knowledge about these suppliers – that 'A' grade from Company X is the same as 'B' grade from Company Y, for example.

This look inside textile sorting factories, and the buyer–supplier chains that connect to them, helps establish the general principles that underpin the recommodification of discard. Four of these stand out. They are: 1) that, as with primary resources, this is an activity grounded in extracting value. Much as with the mining of resources, what matters here is identifying, and then separating out, what is of value (that which can be sold) from what is not (residues and wastes). Processing is central to this. In textile sorting, it is the combination of basic mechanical processing (conveyor belts) with large amounts of human labour that works to identify and extract value from what has been discarded. Given the differential costs of labour across the world, there is scope for multiple rounds of sorting, with each sort eking out more value from the discard. 2) As a process, extracting value in resource recovery is an exercise in material characterization and classification that, in turn, is translated into an economic good. This is where categories and grades come in. The more categories, the finer the degree of material separation. The finer the material separation, the greater the purity of what has been recovered – or, said another way, the less the contamination. But categories need to be translated into marketable goods. This means that they need to be graded and bulked up (here, as bales that are in turn graded). As with all bulking operations, there is the opportunity at this stage for unscrupulous businesses to mix and adulterate product. 3) Simultaneously, the purpose of characterization, classification, sorting, separation and grading is the partial standardization of non-standardized goods. This is what allows processors to sell goods to buyers, but there will always be limits here given that the input is derived from discard, which, by definition (since it is collected from diverse and distributed sources), will always be highly heterogeneous. As such, 4) processors have difficulty guaranteeing quality of supply while buyers work in conditions of asymmetric knowledge. There are, therefore, considerable commercial risks to such operations.

★ ★ ★

In their reliance on large amounts of human labour, textile sorting factories are very much the exception to standard practice in extractive and manufacturing industries in the global North, where – in response to rising costs of labour – the trend over decades has been for increasing levels of automation, to the point where it is now widely anticipated that robots will replace people on the production lines of the future. Those trends are underscored by the nature of much manufacturing activity in the global North. Although there have been some signs of what is called 'reshoring' in recent years, and while the food and drink sector is an important area of manufacture, the primary focus of manufacturing activity in the global North is high end, high value and advanced, comprised of sectors like aeronautics and aerospace, bioengineering, life sciences and robotics. This is significant as regards understanding recent efforts to recommodify discard in the global North, for it means that there is little demand from the manufacturing sector here for many of the low-value goods that currently emerge from processing discard, most especially consumer discard. This has proven particularly critical for the development of recycling in the global North, for what that means is that while materials may be recovered, many are unlikely to be recycled through production processes in the global North. Instead, and as textiles have shown, for much material recovered in the global North to be recycled into more consumer goods, this kind of stuff has needed to find its way to the consumer factories of the world, most of which are in Southeast and South Asia. In this arrangement, recovery and recycling are geographically and economically decoupled. This has had major effects, which I examine in the next section.

Material transformations and the challenges of recommodification

Given high costs of labour and the inevitably capital-intensive nature of much productive activity in the global North, recommodifying discard in this part of the world has been approached almost exclusively in capital-intensive ways. In that regard, textiles recovery is a rare exception. The technologies used thus far have had effects on the shape of the recovery sector that has emerged in the global North, and its outputs. This is because capital-intensive technologies put a premium on processing volume at speed, as the means to recoup investment and turn it to profit. But it is important to recognize that the technologies that were turned to here, and then integrated into recovery factories, also reflect key policy drivers. In other words, policy thinking also helped frame the technological approach that came to prevail in recommodifying discard in the global North. With a regulatory push to rapidly reduce a reliance on landfill, recovering materials for recycling became the favoured and normative approach to managing waste in the global North. Discard got defined as waste, with the weight of discard that is turned into recovered materials serving to demonstrate successful diversion from landfill. The twin imperatives shaping the global North's emerging recovery sector, then, were (and still are) recovery (for recycling) and weight. The capital-intensive processing innovations that recommodify discard in the global North have been oriented, therefore, to technologies that are capable of processing not just large volumes of material continuously and at pace, but also large volumes of weighty material. This favours mechanical technologies and particular materials – glass, paper and card, for example, rather than plastic. Whether those technologies best serve recycling is a rather different question.

Mechanical innovation and the decoupling of recovery from recycling in the global North

Inevitably, the specific mechanical technologies utilized in the recovery sector reflect the discarded things being processed and the main materials being recovered. If we take cars as an example, the primary material to be recovered from them in terms of weight and volume is metal – chiefly steel. But cars are highly complex things, including engines, tyres, batteries, electronics, infotainment systems, upholstery, catalytic converters (containing rare earths), glass, polymers, oils and lubricants as well as metal. It is some of these components and the challenges they pose as wastes that led to the End-of-life Vehicles Directive in the EU in 2000, and the subsequent development of car scrappage (or recovery) facilities within EU member states.[51] Mechanical recovery, however, means that if such things are to be either salvaged or safely removed from scrapped cars, they must be removed at pace (or not at all). It also prioritizes processing volumes of metal at speed in ways that then feed into the metal recovery sector, or the scrap trade. Whereas scrap merchants and dealers have been a feature of economies in the global North for centuries, the addition of car recovery and the high cost of waste management in the global North mean that these recovery activities are not vertically integrated with the scrap trade. In the absence of formalized connections between firms, then, car scrappage is a form of preparation or pre-treatment for the metal recovery business. So, scrapped cars are stripped of certain components and then the metal that comprises the car bodies is crushed to enable efficient transportation. They are then sold on to metals recovery firms, who own the shredding and baling infrastructure that prepares recovered metals for sale into the metal markets. Crushing car bodies may be the most efficient way to recover cars and process them at volume, but in metals recovery terms what it results in is undifferentiated mixed metal, and a lot of lost value. By contrast – and as we saw with the textile sorting factories – it is the scrap metal businesses who capture the value in that recovered metal, by doing just the same sorts of things with metal as the textile sorting factories do with used clothing: extracting value by sorting, segregating and grading these materials, chiefly through mechanized shredding, and then baling them to turn them into standardized goods for sale into the commodity markets.

It was the same story in relation to another highly complex thing that I had the fortune to observe over the years 2009–11, in what was an experiment in 'ship recycling' in the UK.[52] This experiment is worth a brief digression. It shows how recommodifying discard applies to capital goods as well as consumer goods, highlights the challenges involved in recovering value mechanically from complex discard, and shows the limits of mechanical innovation when it occurs in an economy like the UK's.

A brief note on 'ship recycling'

Much as with car recovery, the return of ship recycling to certain economies in the global North for a few years in the first decade of the 2000s owed everything to regulation. Mostly, these days, the major work horses of global trade – cargo vessels, container ships and oil and gas tankers – are dismantled, mostly by hand, on beaches in South Asia; quite why – and with what effect – I will come back to towards the end of the chapter. For the moment, what matters is that this activity has been a poster child for environmental campaign groups, for whom its labour conditions and the environmental degradation it results in has come to exemplify the dark side of globalization. Commercial (or merchant) vessels are, by definition, hard to associate with owners.[53] By the time they are scrapped, the chains of ownership that would connect pollution to specific corporations are even more difficult to establish. By contrast, naval vessels are very hard to disguise, for they are bespoke and clearly commissioned by particular governments. So, when a British ex-naval vessel ended up on Gadani Beach in Pakistan, this was an easy target for environmental campaigners. Named and shamed, the then British government was forced to respond, which it did by developing policy that sought to 'recycle' former naval vessels closer to home, in environmentally contained ways.

Subsequently, an entire class of naval vessels was decommissioned and put out to competitive tender for scrapping. A range of businesses, each working in different locations – in the UK, wider EU and Turkey – and with different (more or less mechanized) methods, successfully bid on the early contracts. But even by the very early stages of the experiment, it was clear that one business (located in Turkey) was way more commercially competitive than the others located in the EU, which, given the 'best value' rules informing UK government procurement, meant that it secured the bulk of the remaining business.[54] The primary reasons for that competitive advantage are readily identifiable: less stringent environmental regulations than within the EU, which allow the facility to operate on a slipway on a beach; lower costs of waste management; lower labour costs; and high levels of demand for recovered metal in Turkey, which is a key global trading hub for steel scrap. But what of the failures – what do they have to tell? Particularly interesting is the comparison that the wider experiment enabled between the value realized from two near-identical 'sister ships' processed simultaneously by two different corporate interests but in very different ways.[55] One of these (a new market entrant) placed a premium on mechanical processing at speed, using an array of heavy equipment that would normally be used in building demolition work to cut the metal as fast as possible; the other (with experience in the business), while using some mechanical processing, approached the recovery process as a scrap metal business, spending considerable amounts of time cutting metal manually (as occurs in South Asia) and then sorting different steels, as well as pulling out more of the most valuable metals such as copper, aluminium and brass. Whereas the first generated large amounts of metal differentiated into basic market categories, the other was able to achieve finer gradations and to segregate more of the most valuable metal, which, because it was vertically integrated, it then shredded and baled and sold into the global commodities markets. Unsurprisingly, it was the latter that realized far more value from pretty much an identical material starting point; and whereas the first left the market, labelling it as impossible to make sufficient money, the other remains in business. The experiment, therefore, is another pointer to the significance of the commodities markets to materials recovery. It also shows that recommodifying discard is about recovering resources, and recovery is only as good as the material quality of what is recovered. When recovery is decoupled technically as well as economically from the markets into which recovered materials are sold, that is when value is well and truly lost.

Household recycling collections

A rather more familiar category of discard than warships is the type of stuff that constitutes regular household recycling collections. It too illustrates the challenges of mechanical processing, showing that these challenges transcend different types of goods and materials. It also serves to reinforce the consequences of the geographical decoupling of recovery activities from recycling, highlighting the dependence of recovery in the global North on global export markets.

On the edges of many of the cities of the global North are giant materials recovery facilities (or MRFs). These are perhaps best thought of as materials segregation factories dedicated to recovering tonnes of material discarded by households, specifically the stuff in 'dry recyclables' collections (think paper, plastic bottles, glass, card, cans). Following collection, this material is first bulked before being transported to the MRF. On arrival at the MRF, material is tipped into holding bays and then loaded onto conveyor belts, where it is pre-sorted by teams of manual labour, whose task is to pull off the belt all the stuff that shouldn't be there.[56] So, in terms of immediate processing, there are parallels with what goes on inside textile sorting factories, but the parallels only go so far, for what goes on subsequently in these factories is entirely mechanized. In an MRF, the focus is on characterizing and separating materials using key material qualities – for example, density and magnetic properties – as well as optical recognition techniques. The technologies work to sequentially separate material streams, so glass, paper from card, certain plastics only (PET and HDPE) and aluminium, from the heterogeneous mass of discard that is delivered to the factory gates.

Mechanical characterization technologies, like other forms of mechanical processing, put a premium on processing volume at speed. But that has had two further effects. One has been to limit the materials recovered to those that can be easily characterized and thus sorted and segregated. In turn, these factories have been designed, and technologies put together inside them, in ways that allow for a fixed sequence of separation and recovery. This makes it much harder to incorporate new materials into the recovery process, or to retrofit them. In such a way, the recovery process has come to be characterized by inertia, or lock-in to a particular array of discarded materials. That inertia is compounded by the inevitable lag between innovation in recovery and wider socio-technological innovation. Designed to process a pattern of consumer discard that prevailed during the 1990s, the most obvious problems facing the current generation of MRFs in the 2020s are: the declining significance of paper in the dry recyclables stream, as an effect of the rapid expansion in digital technologies in everyday consumer lives; the substantial rise in plastic packaging, particularly in relation to sales of food and drink; and the growing amount of card in household collections that is an effect of the inexorable rise of online sales and working from home.

A second effect of material characterization technologies is on the quality of material that is recovered. Indeed, when we compare what goes on in MRFs to the textile sorting factory, we can see that MRF output is broad types of materials (paper, card, glass, aluminium, for example – the equivalent of skirts, blouses, trousers). It is not remotely near the admitted 160 grades of output found in textile sorting factories, let alone the likely number of grades used there.[57] Used textiles are the most finely differentiated of recovered goods that are produced in the global North, but even so, in all materials markets, grades are more than types of material. Paper and card are not just paper and card; plastics are not just particular polymers.[58] And then, grades are crosscut by levels of contamination.

As we have already seen with textiles, grades play a major role in the recommodification of discard, for it is the grades that turn it into traded commodities. Grades also dovetail recovered material with the primary resources sold in the wider commodity markets; so, they work to link the secondary materials market to the primary market. In general, then, precisely because they characterize material type, rather than classify material by grade, the capital-intensive mechanical sorting technologies that are used to sort consumers' 'dry recyclables' generate large amounts of roughly sorted, often heterogeneous – and therefore relatively low-grade – recovered materials. This is exactly the same pattern that we observed with cars and ships. And, as with those activities, the streams of materials that emerge from MRF processing are then baled and sold into the commodity markets. What happens to this output?

Demand for low-grade recovered material varies according to the material and the context. In the global North, aggregates markets absorb much of the low-grade, mixed glass output produced by certain MRFs and there are strong markets for recovered aluminium and PET, both of which are used for product packaging in the food and drinks sector. This means that this recovered material tends to be recommodified in the global North, where it is turned into basic products for path and road construction and cans and receptacles for the food and drink industry, respectively. However, demand for low-grade recovered paper and mixed polymers in the global North is low to non-existent – in the first case because of issues of cross-contamination, and in the second, because of an absence of markets. Low-grade output, therefore, ends up being sold into the global commodity markets and exported into markets where either there is demand for low-grade materials, and/or where cheap labour costs mean that further (hand) sorting can extract more value from it. Inevitably, this means that these materials head out of the global North. Capital-intensive materials recovery in the global North, then, has developed in such a way that it relies on a global export market to absorb much of its output; or, said slightly differently, the material that is recovered from discard through mechanical processing in the global North is reliant on global recycling networks for its recommodification.

Biological and chemical innovation: moving recovery up the value chain

If the 1990s were the decade of innovation in mechanical recovery technologies in the global North, the 2020s are all about the possibilities of innovations in biological and chemical treatments. The latter are often presented as technically superior to mechanical recovery operations, on account of the limited number of times that a particular unit of material can be subjected to mechanical recovery without suffering material degradation. Alternatively, others present chemical and biological operations as complementary to mechanical recovery, with the former tackling what mechanical recovery cannot. Spurred on by 'circular economy' thinking, particular 'problem wastes' have been cast as the starting point for devising novel technological solutions, which ideally will result in a high-value end good. In such a way, so the theory goes, wastes will become resources that can be 'upcycled' into valuable products with a range of end markets. Waste can, in other words, be moved up the value chain. As so often with science-technical innovation, the challenge – as this section shows – is less what can be done technically and rather more in the area of economic practicalities.

At the most basic level, chemical recycling involves using chemical processes (purification, depolymerization, pyrolysis or gasification) either to recapture basic monomers (chemical building blocks) or as the means to convert materials designated wastes into a useful material. The first is a recovery operation that results in material that can be reused in chemical production; the second is a chemical transformation that results in a different recycled material. Biotechnology, by contrast, works with the metabolic qualities of microbes (bacteria, yeasts, fungi). It uses engineered bugs in engineered environments to break down waste materials and/or to transform them into something else. The metabolism of microbes has been used for decades in treating wastewater and producing biogas, but currently chemical and microbial processes are being turned to the problem that is plastic waste.

The scale of the global plastic waste problem is undeniable and urgent, as is the paucity of existing recycling options. A 2016 report[59] highlighted that plastics production increased 20-fold in the 50 years from 1964 to 2014, and is expected to double again by 2034, yet only 14 per cent of plastic packaging was being collected for recycling (compared with 58 per cent of paper and 70–90 per cent of iron and steel), with 90 per cent of plastics being produced from virgin feedstock. Escaping current recycling collection systems, much of today's plastic ends up clogging up urban infrastructure and in the world's oceans, where an estimated 100 million tonnes in 2014 is projected to grow to 1 tonne of plastic for every 3 tonnes of fish by 2025, with plastic anticipated to exceed global fish stocks by 2050. In the face of this challenge, the most progress thus far has been in chemical feedstock recycling, in which monomers are fed back into petrochemical facilities to produce more plastic. The drawback is that such processing applies currently only to PET, which is reduced back to terephthalic acid. This is the area of activity that is furthest down the road to industrial-scale production. Alongside one acquisition, there are already a number of partnership agreements forged between plastics and base chemicals producers located in Europe and European waste management firms, whose role is to supply the petrochemical companies with the plastic feedstock.[60] In such a way, petrochemical companies are following the classic route to recommodifying discard, externalizing the costs of recovery and of materials preparation and/or treatment, and then using partnership agreements to recapture the value latent in this material by incorporating it back into the plastics production process.

By comparison, the road to developing useful materials starting from plastic waste and then to find new applications that can, in turn, be commercialized has been rockier. Aside from the generic challenges that face any innovation process (scale-up, finding appropriate commercial partners and then raising the finance), some of the key challenges facing outputs derived from processing plastic waste chemically are much the same as those facing any processor dealing with waste as a feedstock. These are the inconsistency in the baseline material and resultant challenges of purity and contamination. And, even if the processing can be got to work, there is no guarantee that the output will realize the kinds of applications that might be anticipated from its material qualities. One of the best illustrations of some of these challenges is provided by Plaxx®.

Plaxx is an oil/wax product derived from plastic wastes that are not recycled by MRFs – so plastic films and layered laminates, notably crisp packets. In 2015/16, its developer, Recycling Technologies (based in Swindon in the UK) had got Plaxx and the processing technology (RT7000) to demonstrator status, processing 100 kg/hour of plastic waste into their product. They anticipated getting it nearer to industrial scale (1 tonne/hour) by 2017. At that time, the material qualities of Plaxx were considered to make it ideally suited to marine applications, as its low sulphur content appeared to make it a potential candidate to replace the heavy fuel oils used extensively in global shipping.[61] As such, Plaxx was trumpeted across multiple media platforms; it was seen as a potential saviour of the shipping industry, for whom the reliance on heavy fuels is a major challenge in moving to reduce their carbon footprint.[62] Testing by university and commercial collaborators, however, showed that the wax particles present in Plaxx were likely to cause major problems for engine systems, particularly for fuel pumps and injectors, with research concluding that it was unclear which engines it might work as a fuel for, while also speculating that it might work as a fuel in developing countries.[63] Subsequently, Plaxx has been described as an oil/wax product suitable for use as recovered feedstock in petrochemical processes, to produce new plastic, with the wax fraction open to use as a feedstock for the production of wax blends.[64]

The key technological innovation with Plaxx, however, is not the process but rather the scale of processing. The RT7000 is miniaturized technology, scaled to be suited to small towns[65]; yet, adoption and roll-out have been slow: in 2021 (the time of writing) just one commercial agreement had been signed. In part, slow uptake can be suggested to result from the differences between the business economics of the recovery industry and the petrochemicals industry. Plaxx offers attractive returns compared with the fees charged for waste disposal.[66] However, the business model for petrochemical feedstock recycling favours capital-intensive processing: 30–200 kt per annum would be normal. Modular units, such as the RT7000, can process 2–10 kt per annum (and sometimes even less than 1 kt). This would leave waste management firms (and/or their local authority partners) seeking to sell very small amounts of petrochemical feedstock into the global commodities market, where demand is dominated by the major plastics producers requiring feedstock in sufficiently large and consistent volume to demonstrate a certain level of recycled content.

In comparison to chemical recycling, biotechnological processing of plastic waste is at an earlier stage in the innovation cycle. Most academic reviews, however, are circumspect about the possibilities.[67] Alongside the biochemical challenges, there are the key metabolic engineering challenges, summarized as 'TRY' (titre, rate and yield). Most work, therefore, is no further along the innovation pathway than 'proof of concept' stage, or it has faltered at scale-up, where recovery rates have been too low and incubation periods too long for successful commercial exploitation. Beyond lies the commercial challenge, whereby products derived from expensive processing must compete with products produced from virgin polymers. Ultimately, the contours of success or failure here are determined by oil prices.

An example of microbial 'upcycling' from PET illustrates many of these problems. Recent successes in deriving vanilla flavouring from plastic bottles using genetically engineered bacteria have been trumpeted as the first use of a biological system (in this case engineered *E. coli* bacteria) to generate a high-value product (synthetic vanillin) from a widely available source of plastic waste.[68] Proof of concept, however, while demonstrating technical possibility, does not turn a technical process to an economic good. There is not yet a high-value product here. And vanillin derived from microbes faces considerable challenges in being accepted as an alternative product.[69] Yields are low and incubation times are long, so production is expensive. Even more challenging is the nature of the existing market, within which any new product must compete. While demand for synthetic vanillin is considerable – vanillin is widely used in the food industry, in cosmetics and to make pharmaceuticals, cleaning products and herbicides – the shape of that demand is changing and there are other synthetically derived versions of vanillin that currently dominate global supply. In the food industry, the strong trend since 2015 has been for synthesized vanillin to be replaced by vanilla derived from natural sources, or at the very least sources that can claim to be free of artificial ingredients. That has fuelled global demand for vanilla beans and increased demand for the lignin-derived vanillin by-product produced by Borregard's plant-based biorefinery in Norway. At the same time, demand for natural vanillin has pushed parallel lines of innovation in plant genetics. It is difficult, then, to see any form of synthetic vanillin derived from plastic waste being of interest to this part of the market. Beyond the food industry, any product derived from plastic waste must compete with Solvay's synthesized vanillin. Comprising 85 per cent of the global market, and synthesized from petrochemicals, it currently costs $c.$ $10/kg. That is the benchmark of commerciality currently for synthetic vanillin.

The examples of Plaxx and synthetic vanillin highlight the not inconsiderable economic challenges that face 'upcycling' discard. Indeed, for all the technical possibilities, and regardless of the type of material transformation, there are precious few instances in economic history that demonstrate that the challenges of 'upcycling' recovered materials can be overcome. Shoddy is one; the use of discarded coal tars to synthesize textile dyes – itself a foundation story of synthetic chemistry[70] – is another. Both contain important economic lessons for the present, not least of which is that for discard to become an economic good, treatment and/or processing needs to result in a product that is cheaper than (and of what appears comparable quality to) rival goods. Unless the good itself is open to branding (as can occur with some clothing, for example), distinctiveness and the badge of being made from recovered material will not suffice in the harsh realities of capitalist markets. By contrast, the wider market success of 'downcycled' and closed-loop recycling in parts of the global North is indicative of those core economic principles. Chemical feedstock recycling and the mechanical recovery of PET are the two examples referenced in the chapter, but there are others, notably the use of construction and demolition waste to produce road and path aggregate. But even with this type of recovery activity, there are challenges. Mostly, these illustrate some of the pitfalls of using materials declared to be, and regulated as, wastes as the key input material for product manufacture. A good example of what can happen here is provided by products derived from recycling car tyres.[71]

In 2005, the EU-imposed ban on landfilling car tyres resulted in an urgent need to develop new applications for 'end-of-life' tyres, of which, just over 50 per cent are now recovered for use in newly manufactured goods, with 35 per cent heading in the direction of energy-from-waste facilities. Those tyres that are recovered are first shredded. Mechanical recovery technologies are used to separate metal and fibres from rubber, with the former being sent for briquetting and the latter further granulated. Granulated tyres find a use in asphalt (as infill) as rubber tiles and to provide particles and tiles used in the construction of artificial sports surfaces and children's play areas. In such a way, they have spread all over European urban environments, but there has been a twist in the tale. Granulated tyres, like tyres themselves, are now recognized to be a major source of microplastics in the urban environment, and in 2021, the EU issued a total ban on the future use of granulated rubber, with a 6-year transition period. A critical lesson, then, is that even for those applications that succeed in turning discard to a successful economic product, the regulatory framework can reverse the process, turning a product back to a waste. That is a fate that affects many products, of course, for the history of innovation is paralleled by a history of regulatory interventions that seek to rein back on manufactured toxicity. For products derived from waste, though, regulatory intervention is truly a double-edged sword. On the one hand, it defines the conditions of possibility, for without being declared to be wastes in need of a technical solution, such products would not exist; on the other, environmental knowledge not only continually revisits questions of manufactured toxicity but also lags innovation. In such a way, an application that recovered materials from waste can find itself subsequently being redefined as producing toxicity. The solution becomes a problem and thus is re-rendered a waste. At the very least, we might speculate, such relatively short-term regulatory volatility will act as a brake on investment activity; at worst, it signals that products derived from wastes are a highly risky area of financial investment, not least compared with other potential areas of product development. The corollary is that firms and developers will struggle to secure financial backing and that the commercialization of applications derived from discard will continue to prove challenging.

Recycling futures?

I want to conclude this chapter by turning finally to the intersections of innovation based on wastes, or residues, and regulation. This is not only because this has had a profound effect on shaping the possibilities for recommodifying discard in particular parts of the world, but also because new rounds of regulatory intervention, predicated on projections of increasing resource scarcity, make increasingly normative the need to move economies away from the linear (take-make-dispose) model and towards circular principles.[72] In these formulations, recycling becomes not just a desirable activity but a fulcrum activity, alongside other activities that extend the economic lives of already manufactured things, notably repair and remanufacturing. Recommodifying discard is imagined as an engine in economies, not as a means to cleaning up economies. But just how possible is such an economic future with recycling as its fulcrum, and is it even desirable?

The history of experimenting with materials discarded by industrial processes and/or consumers to generate products is at least as old as industrial capitalism[73], and I began the chapter with one of the most famous such products, shoddy. Although waste is often described as the classic economic externality, there is, then, nothing about capitalist economies per se that means that what is discarded by industry is waste – or stuff of no value. Rather, as 19th- and early 20th-century capitalists recognized, industrial discard is often stuff that's in the wrong place – in a situation (be that a firm or a particular economy) where it is not seen to be of any immediate value.[74] Move it somewhere else and that apparently valueless stuff can be bought and sold and used to make something else – witness the renaissance yarn of shoddy. The irony of much environmental regulation, then, or an unintended consequence, is that in declaring something to be a waste – and insisting that that stuff is managed as a waste, which often means sequestering that stuff as a waste – it makes it far harder to do something with it that might enable it to be recommodified. The first sections of this chapter made clear this is why so much discard is moved out of the global North to be recommodified. It is precisely because the emergent economies and economies of the global South provide the economic and (lax) regulatory conditions in which discard can be revalorized that trade flows in this way.

Unsurprisingly, those conditions have led to the emergence of agglomerations in the global South in which local economies rest on the circular economy trinity of repair, remanufacturing and recycling. They provide an insight into the kind of economies that emerge when recycling becomes a fulcrum activity. An exemplar case is Bhatiary in Bangladesh.[75] Bhatiary is the largely unacknowledged economic engine that sits behind the headlines that are captured – for all the wrong reasons – by the ship-breaking activity that occurs on the beach near Chittagong. Justifiably, this activity is widely slated internationally for its labour conditions and for its environmental pollution. Yet, notwithstanding protracted opposition, it continues, as it does at Alang in India and Gadani in Pakistan. And it does so not just because ships need to be scrapped once they reach the end of their economic working lives, but because ships (and particularly large ships with lots of space that are easy to cut up, like oil tankers) are a plentiful source of ferrous scrap.

Steel production is foundational to modern economies. It is the basis for most construction as well as of critical importance for huge swathes of manufacturing activity, but for economies lacking the basic raw materials (coal and iron ore) that allowed the industry to develop first in Northern Europe, the route to primary steel production has been via the electric arc process. This process requires very large amounts of ferrous scrap metal. For economies like Bangladesh's, the route to domestic steel production has been tortuous and slow. In this particular case, it was bound up in the political ramifications of India's partition and then the war with, and separation from, Pakistan. Indeed, it was the inability of Bangladesh's one primary steel plant (located in Chittagong) to supply the increasing needs of the Bangladesh economy that led to the development of secondary processing capacity, vertically linking ship-breaking on the beaches with rerolling mills just inland producing rebar rods for the construction market. But ships provide more than just ferrous scrap; they are designed and built as mini cities, for they must sustain human life for long periods away from land. So, while ferrous scrap is the primary material that comes off the Chittagong beaches, it is accompanied by power equipment (motors, generators, engines) and their component parts, a vast array of interior construction boards, furniture and furnishings, wiring and cabling, electronics, and all manner of domestic infrastructural goods – toilets, sinks, basins, taps, laundry devices and cooking utensils. These goods, which are auctioned off by ship-breaking firms to firms specializing in repair, reconditioning and remanufacturing, then find their way into the domestic market – either as capital goods for firms producing primarily for the domestic market or as consumer goods serving the Bangladeshi consumer market.[76]

The grounding of the Bhatiary complex in one of the dirtiest of global industries is certainly not to be celebrated, but it provides an object lesson in what is required economically if recycling is to become a fulcrum activity, and in the perils of this course. It would not be too much of an exaggeration to say that Bhatiary only developed in this way out of conditions of economic failure and fragility. It was the failure of the primary steel industry to keep up with demand in an economy developing from a very low base, combined with the vulnerability of the Bangladeshi economy internationally, that provided the initial impetus for agglomerative conditions to emerge. But what has developed is an entire industrial district that is totally reliant upon the continued supply of ships to the beaches. That faces a sustained double-pronged attack, from non-governmental organizations and the regulatory community on the one hand, for whom dirty ship-breaking is an activity that either needs to be banned or cleaned up, and from new primary steel producers on the other, for whom (in a direct echo of the shoddy tale) secondary steel is an uncertified, dodgy product – a particularly potent label in a country where building collapse is a not uncommon occurrence. Yet, if the ships dry up, the entire local economy and beyond would collapse, for the tentacles of this activity are distributed across Bangladesh, spilling over to affect not just construction but domestic manufacturing, retail and consumer markets, and economic growth. Bhatiary, then, provides a lesson in economic lock-in and an illustration of the extreme dependency that happens when a particular recycling activity drives an economy.

It is important to recognize that the argument to recast the economies of the global North through circular principles, and with recycling at its heart, is not an argument to remake economies in the image of Bhatiary – even though the same impetus of resource scarcity is the primary motivation. For one, the economies of the global North are advanced economies, the success of which has rested and continues to rest on the restricted and constrained development of other countries, like Bangladesh. They are also economies where economic activity is more tightly regulated. Further, the impetus to circularity is coming from regulatory intervention, not from organic evolution and the emulation that results in. So, it is unlikely that entire clusters or industrial districts, grounded in such activities, will emerge. Rather, the greater likelihood is that regulation will drive manufacturers and retailers to use more and more recovered material. Stated another way, policy levers will be pulled that increase the supply of secondary materials and drive demand for such materials. What that amounts to, I contend, is the 'shoddification' of manufacture. It is to extend the same principles that occurred in relation to the innovation that was shoddy itself (eking out resources in conditions of resource scarcity) into multiple sectors and multiple firms. Just how possible, or even desirable, is this?

There are at least three key challenges that I can see with this approach. First, there is the unevenness of the potential for shoddification across the manufacturing base. So, while increasing the content of recovered material in, say, packaging materials makes a very great deal of sense, given that these are highly transient goods whose purpose is to enable the logistical movement of goods between manufacturers, retailers and consumers, it is way more challenging when what is being manufactured is a high-value, precision-engineered good or component. In circumstances where what is being produced is a high-value good, only recovered materials that can be certified to the same standards as their primary counterparts will pass the acceptance criteria thresholds for advanced manufacturing.

Second, there is the dependence of shoddification on the blending of primary raw materials with secondary recovered materials. This is a scaled-up version of the problem that confronted Benjamin Law, and it can occur either at the level of the material itself and/or in the manufacturing process. Outside of closed-loop production processes and contracted agreements between firms with access to large volumes of discard (chiefly the waste management majors) and baseline manufacturers such as the petrochemicals industry, heightened demand for recovered materials will push for greater market concentration on the supply side, favouring those who can supply large order books. Small traders based in the global South will be unable to operate at this scale. Instead, markets for at least some recovered materials are likely to become increasingly concentrated and to move more within the orbit of the global commodity traders. They are the key intermediaries in the primary global resource trade. As recent work has shown, however, there have been a not insignificant number of high-profile instances of adulteration, mixing and fraudulent mislabelling of goods, involving the blending of primary resources (particularly oil) by these traders[77], which are suggestive as to how control of the resource trade and blending can be used for financial gain. Recovered materials are even more open to such forms of blending, since they are often produced from dispersed geographical areas and then consolidated and amalgamated at a central location. Then there is the issue of blend control and its relationship to manufacture. Benjamin Law devoted huge amounts of time trying to perfect machinery that could cope with a blended material, and different blends of primary and shoddy fibres, but tellingly, that was in an agglomerative situation that provided the ideal circumstances for experimentation, and in a corporate context in which the manufacturer controlled the blending process and had first-hand knowledge of wool fibres. In a scaled-up, open market in which the regulatory impetus is merely to increase the amount of recovered content, such intimate technical knowledge of how materials relate to processing technologies is less likely, and the potential for control is far more diffuse. Further, doubts will persist over the qualification of recovered materials. So, one can imagine a future here where only the least risky of recovered materials might be incorporated into manufactured goods, which basically boils down to metals, some polymers and packaging.

Thirdly, there is the problem that beset shoddy itself and that has persisted through the long history of shoddy goods: its lack of durability. While blending primary and recovered materials ekes out resources per se, its result in an economic good is generally recognized to be of inferior material quality, which has an inevitable effect on the lifespan of that good. A key question is how significant that is. The answer to that question varies according to the nature of the good. If the good is transient, that inferiority is generally not problematic. A case in point is packaging, but there are other examples of paper goods that include varying proportions of recovered content – toilet rolls, kitchen rolls, newspaper and writing paper. Softness, absorbency, clarity of print definition and the aesthetic experience, look and feel of virgin paper all matter here, showing that there are material limits to how much recovered content will sell, even with paper, which is not generally sold for its durability. If a good is sold on its durability, then there are likely considerable limits on how much recovered content might be included without detriment. For such goods, the circular solution is more likely to be enhancing their capacity for repair rather than engaging in the false promise of recycling. Shoddy, lest we forget, is a product that taught lessons. It may have lined capitalists' pockets but it resulted in goods that quickly fell apart – and it still does fall apart, as Panipat's blankets show. For all the worthy intentions that lie behind it, shoddification is both limited in its potential and carries many of the same 'race to the bottom' risks as its product predecessor.

6

Waste, Money and Finance

Or, how turning discard into waste turns waste into an energy resource and an asset

In this chapter, I examine what happens when discard becomes waste. My argument here is that while there is no inevitability that discard will become waste, there are strong pressures that it does. More than this, I argue that there are drivers that work to ensure that the proportion of discard that is captured as waste actually increases. Rather than this being a period characterized by waste reduction, the world is currently in a situation that demands the converse: that more waste is generated.

To make this argument is to go against an industry and regulatory narrative that presents the current situation as one in which business and regulators are increasingly aligned in their pursuit of Zero Waste.[1] That, I maintain, is 'greenwash', for – leaving aside the impossibility of that goal – what the Zero Waste narrative does is to mask that the waste management business (or, as it increasingly casts itself, the waste-to-resource business) is very much a waste *business*. In other words, it makes (very large amounts of) money by attending to materials classified as waste. So, at a fundamental level there is every incentive for the waste business to want more waste. But just what kind of 'more waste' does it want?

Of critical importance here is the category of waste that the industry labels as 'residual' – or that which is not suitable for recovery for recycling, or reuse. Taking the materials that can be recovered for recycling, sorting and separating them, and then selling them into the commodity markets is one means by which the waste industry makes money from the waste stream. That was the focus of Chapter 5. But the other – and much bigger – source of revenue, at least according to company accounts, comes from the management of the residual waste stream. To understand why residual waste generates the most revenue for waste businesses, we need to understand the technologies that are used to attend to it, but not just as technologies, which is how the waste business presents the management of waste to outsiders. Here, emphasis is placed on the technical processes that enable the safe treatment of material declared to be waste. Instead, I focus in this chapter primarily on how these technologies turn residual waste to money. Turning attention in this direction shows that there is indeed every incentive for the waste business to want more waste.

As important is that the incentives to make more waste – or, to specify this more tightly, to capture more material as waste – go well beyond the waste business. This is because residual waste has proven a fertile terrain for finance capital. It therefore sits squarely within the major transformation of late 20th century and 21st-century capitalism, namely the shift to financialize pretty much everything.[2] How this works goes something like this, connecting what at first sight look to be very different sectors and activities: residual waste, water, infrastructural services such as roads, railways, airports and ports, and other services such as higher education, healthcare and elder care. Connecting these seemingly disparate sectors is their capacity to be turned to assets[3], and hence converted into a revenue stream. This occurs through capitalization – or processes and practices of valuation that shape an expected future monetary return on making an investment in that asset compared with another.[4] What makes all these sectors particularly fertile terrain for assetization is the pretty much guaranteed continued and/or expanding demand for their services. That demand is based on population dynamics and – in varying measures – increasing levels of urbanization and mobility, as well as patterns of consumption that define an increasingly global middle class (Chapter 3). In the case of residual waste, the inevitability of the association between a growing global population, increasing levels of consumption and increasing volumes of discard joins with the necessity for that discard to be managed, through collection and treatment, and for the majority of that discard to require treatment through residual waste technologies.[5] Populations (typically of cities and local municipal areas), and the number of dwelling units within them, become the basis for an estimated yield of residual waste; projected population and housing growth, and data about the amount of residual waste (based on collection data), becomes the basis for future residual waste projections (or yields) over decades. The infrastructure for dealing with this present and future residual waste, along with the waste it handles and treats, is then opened up to financial techniques that turn it into an investment product. Offering returns over decades, residual waste turns out to be portfolio ballast for investors – the kind of stuff on which institutional investors such as banks, private equity firms and pension funds rely.

The chapter proceeds through two main sections. The first provides a long historical view of global waste regimes in comparative perspective. It highlights their beginnings in epidemiology and public health in 19th-century Europe;

the connection this had to sanitation, particularly in an urban context; and the emergence of a collect-and-dispose approach to managing waste. At the heart of this is the competition between two technologies, which persists today, in form if not in technical details: incineration and landfill – or burning and dumping. That regime held sway for much of the 20th century, with the exception of the period of the two world wars and their immediate aftermath, during which appeals to salvage and recovery figured strongly.[6] But by the turn of the millennium, the collect-dispose regime had begun to be supplanted by a waste-to-resource regime. The impetus for that transition initially came from climate science, specifically the growing evidence for the potency of methane as a GHG, and landfill's contribution, as a methane generator, to GHG emissions. As such, it was in the part of the world where climate science held most sway at the time (Europe) that the transition to a waste-to-resource regime first began to take shape, with the form of that regime coalescing around the identification of landfill as a technology of last resort. By contrast, in other parts of the world, and in response to what is largely a World Bank narrative[7], the emergence of a waste-to-resource regime has been linked more to a perceived need to reduce the volume of waste associated with rapid economic and urban growth. It has been about trying to reduce the size of waste mountains. As I show, through a focus on Europe, China and India, there is no single dominant version of what a waste-to-resource regime looks like. But what is increasingly clear is that incineration, in the form of energy from waste, is becoming the dominant technological solution for managing the residual wastes of the world's major cities. This is significant because it means that large amounts of any city's wastes will become classified as residual waste – they have to be if they are to be handled via energy-from-waste technologies. There is a need to be clear about what this represents: it is the means to laying the foundations for much of the world's waste to be captured by big capital.

In the second main section, I turn to examine the range of corporate interests involved in treating and managing these residual wastes. They are transnational firms specializing in energy-from-waste technology, plant design and construction or plant operation. Most are headquartered in the global North, or China. Winning tenders for energy-from-waste plants is argued to give exclusive rights to the wastes generated by a municipality. It is a means of enclosure: capturing wastes as private property for the purpose of extracting economic rent from their treatment. This is why energy-from-waste plants offer huge scope for financial investment. Correspondingly, in the final part of this section I examine how energy from waste has been captured by finance capital and how it is turned into a financial investment opportunity; and I establish that waste is increasingly a means to wealth generation and accumulation. First, though, there is a need to consider what is meant by residual waste.

A brief note on residual waste

The term 'residual waste' is a recent addition to the waste management lexicon. Its arrival coincides with the introduction or, more correctly, re-introduction of thinking about waste as a resource. So it is that 'residual' became the term reserved for that fraction of the municipal waste stream that is left over once other materials have been recovered for recycling or redirected into reuse. In Europe, this came about in the 1990s, and it was a key moment.

Differentiating residual waste within the municipal waste stream was both a political and a technical intervention. European environmental policy at this time had recognized the contribution of landfill to GHGs through methane emissions that result from the decomposition of biogenic materials. At the same time, the EU was also developing a suite of policies relating to water quality, including groundwater, in which the potential deleterious effect of leachate from landfill was acknowledged. As a result, landfill became the bête noir of European waste management policy. It was labelled as a 'disposal' technology – or a technology of last resort, to be avoided if at all possible. That view was enforced first through the imposition of a landfill tax, which introduced progressively higher charges on managing waste through landfill, and second through the concept of the 'waste hierarchy'. The latter is a mnemonic that lays out preferential ways of treating wastes. At the apex of the hierarchy is waste avoidance and prevention – practices that, since they avoid discarding material and categorizing it as waste, effectively bypass waste management. Below that is reuse, then below that recovery for recycling, followed by options for managing residual waste. All versions of the waste hierarchy place landfill at the very bottom of the hierarchy. But above this, and below recycling, was a category that needed to be filled technically as well as politically, not least because there have long been arguments that incineration (the primary technological alternative to landfill) is just as much a disposal technology as landfill. In other words, it was politically necessary to demonstrate a difference between other residual waste management technologies and landfill. So it is that recovery activities have become core to residual waste management technologies, with measuring recovery in residual waste management technologies a critical means to qualifying their difference from, and superiority to, disposal technologies.

In practice, what that means is that incineration technologies have been reconfigured as energy-from-waste technologies, generating electricity that can be fed into the grid and/or supplying heat through district heating networks. The biogenic fraction of municipal waste (effectively food waste) qualifies these plants as generating renewable energy, while the 'R1' designation (a formula that calculates efficiency in energy generation) qualifies which plants are waste recovery facilities. In the EU internal market this makes them eligible to import waste from elsewhere in the EU.[8]

This short technical summary opens the door to a much wider transformation in municipal waste management, one which is being enacted globally but at different speeds and with different intensities. This is the shift from landfill to incineration – or, more precisely, incineration as energy from waste. It also foregrounds the interplay between landfill and incineration as waste management technologies. There is a long backstory to this. The next section takes the long view, charting the ebb and flow between these technologies first in Europe, and then in contemporary global perspective. It also highlights that money and finance is never far away in accounting for these shifts.

Waste regimes in historical and global perspective

The narrative that dominates contemporary waste management is the one introduced in the previous section. It centres the shift from landfill to incineration as energy from waste, or from disposal to recovery. As ever with short timelines, this story obscures much. Indeed, if the temporal lens is widened, the history of waste management in Europe and the US tells precisely the reverse story – of a shift from incineration to landfill. To help understand these shifts, it is useful to turn to the concept of waste regimes.[9]

Broadly speaking, a regime (any regime) is a means to societal ordering and it typically will involve political as well as social, economic and cultural forces. Think of how the term has been used to describe particular states (for example, Iraq and Afghanistan) under certain political leaders (Saddam Hussein and the Taliban), or how it is applied to major institutions and organizations, from firms to football clubs to news media outlets, under different leaders and leadership teams. A waste regime, then, is a means to the societal ordering of waste. It requires, prima facie, that discard be identified and defined as waste. But it will also order waste in four key ways: 1) through political ideology and organization (for example, commitment to the 'free market' v state regulation, authoritarian v democratic states) and political processes, notably policy interventions; 2) technically, through the technological possibilities that are favoured; 3) socially and culturally, in the sense of how waste is regarded and valued (or not); and 4) materially, in that the wastes societies generate reflect their patterns of consumption while at the same time being an effect of the technologies that are used to attend to their discard.

There is a lot that is useful in the concept of waste regimes, and I draw on much of this in what follows – in particular, the emphasis it places on thinking about how waste is produced politically, technically and materially. But, like all regime thinking, waste regimes have an inbuilt tendency to encourage sequential thinking and to highlight transitions, from one arrangement to another. So, even though it produces a longer view, the result is a linear narrative all the same. Linear narratives can be problematic. As I show in this section, the broad oscillatory trajectory produced by the long view – of 'incineration-landfill-incineration' – masks a far more complex picture – of the co-existence and competition between these technologies. So, while it is accurate to see particular periods as dominated by one (or other) of these technologies, it is not strictly true to say that one comes to supplant the other. Rather, particular periods provide the circumstances in which, for a variety of reasons, one is favoured over another and thus becomes the dominant mode of (residual) waste management. As we will see, at the heart of this is money – something that narratives of the history of waste management are remarkably coy about.

The story begins with the foundation of something recognizable as waste management in the global North.

Waste, public health and sanitary disposal: the emergence of a waste collection-disposal regime

From the latter half of the 19th century, the association between waste and ill health was recognized by epidemiologists and public health specialists in the then major global cities (London, Paris, New York), as was the association between unclean water supplies, ill health, disease and deaths. As scientific knowledge increased, dumps, dead animals lying in the gutters (chiefly fallen carriage horses but also domesticated animals) and matter discarded into streets were all identified as key disease vectors, through their connection with vermin (rats), flies and, eventually, bacteria. Rapidly growing urban centres with dense overcrowded living conditions were readily identified as petri dishes for killer diseases such as tuberculosis, cholera and measles, and the resultant ill health of urban populations was acknowledged to be a key social and political problem. The result was the identification of a new branch of health: public health.

The new profession, through its sanitary inspectors, became closely associated with putting in place the conditions for a cleaner, more sanitary mode of urban life. Their work connected strongly with the emergent turn to modernizing cities.[10] Those connections are at their clearest in three areas: the creation of clean drinking water supplies, through the separation of 'clean' (or potable) and 'waste' water; the attendant creation and/or radical improvement and extension of sewer systems to remove human wastes (and the development of sewage works to treat those wastes); and the removal of discarded matter from the streets, be that dust, organic matter, textiles or metal. These changes were encoded in a variety of public health acts that, across Europe and then the US, endeavoured to create cleaner, safer cities.

They recognized something that has come back to bite more recently in the COVID-19 pandemic: that, if they are to thrive, cities' economies require healthy populations, a pre-determinant of which is sanitary living conditions.

Across Europe and the US, the public health acts of the late 19th and early 20th centuries laid the responsibility for this programme of works with local government, or municipalities. They also firmly established the connection between discard and the category 'waste' and between waste and waste management. Seen through a public health lens, discard got redefined as waste – or troublesome, human life-threatening, valueless matter. It was also deemed essential for public health improvement that those wastes were appropriately managed. So it was that waste came to be seen as, and defined in terms of, its management. Unpacking that still further, what that meant was that wastes were seen as: 1) needing to be routinely removed from homes, streets and public places, hence the advent of municipally organized household collections and cleansing services, for streets and public spaces; and 2) once collected, they needed to be disposed of in a manner that was deemed to be sanitary. A regime of municipal waste management, therefore, had emerged, one defined in terms of collection and disposal – two distinct yet connected practices that were then translated into political responsibilities.[11]

Running across this, and underpinning the entire regime, was the sanitary imperative. The latter shaped social and cultural values, turning waste into stuff that had to be separated and removed from urban living and rendered 'out of sight and out of mind'. It also affected technologies of disposal. For public health professionals, dumps were the antithesis of sanitary disposal. Although they removed wastes from urban homes and streets, they did so by displacing the problem, resorting to geography to solve a public health problem and often concentrating that problem in a highly visible way somewhere else, typically in impoverished neighbourhoods on city edges where public health concerns were already high. So for the public health inspectors of the time, incineration (or, controlled burning) became the normative technological mode of dealing with a city's waste. Burning was seen as the clean, sanitary way of attending to waste. Yet – and this is the blind spot of thinking through regimes – dumping remained widespread, as can be seen from the case of England.

Early 20th century reports from England show that while 221 local authorities had incinerators, 908 did not, and that 709 towns continued to use dumps as their primary means of dealing with waste. They also show that 50 per cent of the largest towns and cities and 25 per cent of all local authorities exported waste outside their local administrative area, typically to adjacent, more rural areas.[12] There, it was dumped in open, uncontrolled conditions on land that was declared to be 'waste' or 'derelict', or in gravel pits and quarries. Contemporary descriptive notes of the South Hornchurch dumps in Essex by J.C. Dawes from the Ministry of Health Inspectorate provide a graphic illustration: '[There is a] scarred and fissured surface ... extending inwards for a considerable depth from the tipping face; evidence of extensive and deep-seated fire which frequently ... reaches up to the surface and envelops the great malodorous mass in a characteristically evil smelling smoke.'[13]

Quite why dumping carried on in tandem with incineration is unclear. Speculatively, it might be suggested that the early incinerators (known as 'dust destructors') were small scale and thus would have struggled to cope with the expanding volume of waste being collected. But to my mind, the more likely explanation is that even at this point, the relative costs of incineration versus dumping were having an effect on how wastes were managed. Incineration – even when small scale – involved higher capital costs than landfill. Boilers, furnaces, flues and chimneys, and the buildings to contain 'controlled burning' all needed to be purchased, manufactured and constructed. Then there were the operating costs: the furnace and boilers of early incinerators relied on additional coal-firing and a continual supply of coal was required; routine daily maintenance was essential to ensure safe operation; furnaces needed labour; and the ash that resulted still needed to be managed (either by dumping or by finding a market or use for it, typically aggregate). So, the capital costs of incineration will have greatly exceeded those of dumping, which entailed no more than the purchase of an appropriate area of land. The same would have been true of operating costs, with dump operations requiring little other than basic manual labour. Capital and operating costs impinge on 'gate fees' – or, the charge that the owner of a waste management facility makes for accepting waste into that facility. A broad rule of thumb is the lower the costs, the lower the gate fee, and the more competitive the technology in terms of the price that is charged to a consignor to discharge waste material. So at this point in time, dumping was far cheaper than incineration, meaning that municipalities could incinerate some of their wastes while simultaneously saving money by continuing to export large amounts of it to dumps outside of their administrative area.

If money, in the form of cost, lies at the heart of the persistence of dumping alongside incineration, it is also there in the decline of incineration and rise of landfill through the 20th century. Sticking with what happened in England helps to elaborate.[14]

Reimagining the dump: England's turn to landfill

After World War I, local government budgets in England came under increasing strain as a result of the deepening economic depression that affected Europe and the US. Block grant budgets were reduced by central government, core services came under pressure and, for local government officers, controlling the (high) costs of urban cleansing became an increasing priority. In this changed financial context, incineration came to be seen as an expensive technology. Cheaper alternatives needed to be found – and urgently – for managing England's wastes. Although there was some talk of the merits of recycling and salvage, the nascent field of waste management saw little alternative but to revisit what they called 'tipping'. What transpired was a reimagining and reworking of the dump through technical practices of waste management. Tipping transitioned to become 'controlled tipping', or what is now known as 'sanitary landfill'.

Dumping is, as the term suggests, literally dumping. Matter is deposited where it is dropped. Over time, layers accrue as mounds that become progressively higher, hence the contemporary term 'waste mountains'. Archaeologists call these dumps 'middens'. The new field of waste management, however, positioned dumps within a technical language. Here, dumping became recast more narrowly, through different practices of tipping. Tipping is what happens when wastes are delivered to waste management facilities. It is literally what occurs when loads arrive and are off-loaded from vehicles, and the phrase is still in use in the reception areas of contemporary state-of-the art incineration facilities as 'tipping floors' or the 'tipping bays' that feature in the reception areas of materials recovery facilities. The use of the term 'tipping' shows the close links between the waste management industry and the haulage industry; it is the connection point between the technologies of collection and disposal. But, in a key twist, the waste management industry then redefined dumps as 'uncontrolled tipping', in so doing setting up a binary that was critical to the reincarnation of these methods of disposal and foundational to the rise of the sanitary landfill.

Unlike uncontrolled tipping, controlled tipping – or sanitary landfilling – makes use of naturally occurring or engineered barriers and caps to separate off and manage materials declared to be wastes. It also more tightly orchestrates the work of 'tipping', shifting this from the loose temporal and geographical ordering that occurs in uncontrolled tipping to a more carefully choreographed practice, whereby loads delivered are tipped in a pre-determined area in prepared 'cells' or depressions and then covered with earth and compressed to enable sequential layering. The entirety of the controlled tip is lined with a barrier membrane (in the early days, earth and clay and subsoils, now engineered, geotechnical membranes) to stop the contamination of groundwater. When declared to be 'full', the landfill is capped, that is, it is covered over and revegetated. So there is a future amenity value to controlled tipping, which is also a means to forgetting, as what was once a site of waste management becomes a park, grazing land, woodland, with the only trace of its past (if it is a recent landfill) being the collection of methane gases.[15] So, and in response to political necessity, waste management professionals were able to work with the technical science of containment to convince municipalities that the controlled tip was a safe waste management practice and to imagine a future for that land beyond that of waste management. This is a winning political formula, one that has been applied to locally unwanted land uses in many other spheres over the years, from opencast mining to attempts to site nuclear waste repositories. But what it meant was that the door was open for landfill to take over as the dominant mode of managing waste. By the 1960s in England, landfill accounted for handling 90 per cent of waste.

In many parts of the world, as in England, landfills and/or dumps were the 20th century's waste management technology of choice. It is not too much to say that this was a global waste regime, albeit with differences in the manner of collection and in the techniques of disposal (mechanized or manual). Those differences reflect political differences, levels of economic development and the relative prominence given by different countries to public health and environmental care. In many parts of the world, the absence of a strongly developed local state and allied levels of municipal governance, together with weak public health directives, has meant that waste collection was left to the market, in the form of the informal sector. The consequence is that only those wastes that have local market value were collected. Across the world, these conditions have favoured the waste management technologies that were able to dispose of it most cheaply – landfills, of one form or another. Where conditions have varied are in the degree to which public health concerns allied with concerns over environmental degradation to deem sanitary landfills a necessity. This is how sanitary landfills came to prevail in the global North, whereas uncontrolled dumps were the norm everywhere else.

In the 21st century, landfill remains the dominant mode of waste management globally, but it is widely seen as the least acceptable technology for dealing with waste. That change is, in turn, indicative of a broader transformation: the transition from a waste collection-disposal regime to a waste-to-resource regime. However, the transition to that regime globally is highly uneven, while what that regime entails is contested. Some parts of the world have barely begun that transformation, while others have travelled a long way along that road. So, what does this new regime look like and how does it vary across the world?

The turn to a waste-to-resource regime

Let's begin in a part of the world that might be anticipated to be cutting edge: in the US. In the US, right through the latter half of the 20th century and in line with the highest levels of consumption in the world (Chapter 3), the major cities vied for the dubious accolade of having the world's largest landfill. Key contenders for that title include Fresh Kills on Staten Island, which opened in 1948 and closed in 2001. At its peak, in 1986, Fresh Kills was handling 29 kt of waste a day – just from New York. Out west, Fresno in California opened in 1937 and is the first instance of a sanitary landfill (or controlled tip) in the US. It closed in 1989 and is now a National Historic Landmark, an indication of the importance of landfill to 20th century urban life in the US.[16] Puente Hills on the edge of Los Angeles also has claims to being the largest, with mounds 150 m from the ground, and in 2012 (the year before its closure), 1,500 trucks per day were delivering 12 kt of waste from LA.[17] Today, the US remains highly dependent on landfill as the means to manage its wastes, with 54 per cent of municipal waste being handled through this technology.[18] With relatively low recycling rates (24 per cent) and with roughly similar levels being processed by a new generation of energy-from-waste incineration plants[19], the US remains firmly a waste collection-disposal regime. Its lack of transition to a waste-to-resource regime is widely seen to result from the lack of purchase of environmental thinking in the federal policy domain, combined with widespread opposition to energy from waste as a technology. In such a way, the conditions for the perpetuation of landfill have been maintained, while the global export markets for recyclable materials discussed in Chapter 5 are the primary means of diverting US trash from landfill.[20]

Elsewhere in the global North, landfill has been under a sustained attack. The kernel to that attack has, once again, been through scientific progress encoded in scientific knowledge, this time not from epidemiologists but from climate scientists. Landfill's role as a methane generator, and thus as a contributor to GHGs, was the catalyst to acknowledging that while it might have removed wastes out of sight and out of mind, this was a dirty, polluting technology. In line with the prominence given to climate science there, the EU bloc provides the most developed example of anti-landfill thinking and policy development in the global North, and it is the area that has advanced most within the global North in the transition to a waste-to-resource regime.

As we have already seen, the key policy driver in the EU came in the form of a landfill tax allied to the key mnemonic of the waste hierarchy. The two policies were central planks of the Landfill Directive, which came into European law in 1999. At the level of member states, the tax applied to businesses and municipalities alike and it introduced a stepped and increasingly onerous charge on the management of wastes through landfill, placing the burden of taxation on the generator of the wastes, effectively raising the cost of using this technology, to the extent that other waste management technologies were able to become economically competitive. With a ceiling of £80/tonne, the tax has proven mind-concentrating for businesses generating high volumes of waste as well as for municipalities, which, while not strictly speaking generators, have the political responsibility for collecting wastes and then disposing of them. Subsequently, and in line with its circular economy goals, the EU has intensified and concentrated its interventions, stipulating that in addition to the continued operation of the landfill tax, no more than 10 per cent of municipal waste should be landfilled by 2035 and that no recyclable, recoverable or energy-generating material is to be sent to landfill from 2030.[21] So, not only is the volume (and weight) of material heading to landfill within the EU being severely reduced, but the material qualities of wastes being permitted into landfills is being restricted to non-GHG generative materials. Significantly, that does not erase landfill as a technology. Rather, it continues to recognize the need for landfill – but as a destination of last resort for hazardous, non-hazardous and inert materials.

There are two primary strands to how residual waste management technologies have evolved in response to the emerging regime in the EU. One has been to return to incineration – or, more accurately, to reimagine and rework a technology that had never gone away but that had fallen out of political favour in uneven ways across Europe, often as a direct result of anti-incineration campaigns.[22] In parts of the EU (notably Scandinavia and Germany), incineration linked to district heating networks has long been an important means to managing wastes. Such arrangements were commonplace in many cities through the 20th century – quite unlike the prevailing landfill practices of Southern and Eastern Europe and the Western periphery (the UK and Ireland). Modern incinerators can generate electricity as well as provide heating. Along with modern flues and chimney systems, this has proven critical to the incineration renaissance in Europe. Pan-European market incentives for the generation of renewable forms of electricity meant that incineration was able to be rebadged as 'energy from waste', its status as 'renewable' resting on saving biogenic waste from being landfilled and therefore from generating GHGs. Waste-burning plants, therefore, have become power plants, with waste being reimagined as a clean, non-carbonized energy resource – a sleight of hand that conveniently sidesteps the embodied carbon in the waste that is burned, and a continued dependence on that waste as fuel.

Rather different is a second strand of technologies that seek to work with the organic fraction of residual waste (chiefly food waste), and to turn that to a resource. These are newer technologies, certainly when compared with incineration. So, they have had the usual difficult ride – of technical scale-up and attracting finance (Chapter 5). Whereas one of these technologies (anaerobic digestion) can operate in tandem with energy from waste, the other (mechanical and biological treatment; MBT) is a direct competitor to it. The first works exclusively with the food waste fraction of residual waste. It takes that waste (collected at source from households and businesses in separate food waste collections) and processes it in anaerobic conditions inside a digester, to produce a compost organic fraction and a biogas. As with electricity derived from energy-from-waste plants, this biogas qualifies as a renewable energy source, supplied this time to the gas grid. The second technology (MBT) takes the entire residual waste fraction and then seeks to extract more recyclable value (paper and metals, for example) from it through both mechanical and biological processes – respectively, separation and shredding and anaerobic digestion. The result of the biological processing, much as in a standard digester, is a compost-like material along with a biogas. At an early stage in the technology's scale-up, it was anticipated that this compost-like material would be open to a whole host of markets and uses. That, however, has not transpired, due to the highly heterogeneous nature of the input material. Unlike food waste separated at source, this organic fraction arrives at plants mixed with everything else in the residual waste stream. It therefore needs to be separated out at pace. This is not easy to achieve technically, and cross-contamination is pretty much inevitable. It is this close association with other wastes that has disqualified MBT-derived compost as a compost product, leaving it open, at best, to non-market uses, for example as a soil improver in areas that are being restored and regenerated, through reforestation and tree planting for instance; at worst, if no use can be found, the material has to be landfilled. If it is so configured, an MBT plant can also produce an output called 'refuse-derived fuel'. This derives from mechanical operations and is suitable for burning in some energy-from-waste plants, incinerators and multi-fuel power plants.

In countries where there has been strong public opposition to energy from waste (such as England), mechanical and biological treatment proved politically attractive. It seemed to offer the win–win option of enhanced recycling minus incineration. As such, a substantial minority of England's municipalities commissioned such facilities as the means to managing their residual waste fraction. Unfortunately, the promise vastly exceeded the subsequent operational reality, with many of these plants facing a litany of technical, operating and regulatory challenges, leaving municipalities with a very large problem of needing to find alternative solutions to manage their residual waste.[23]

The broad contours of a waste-to-resource regime are now firmly established in the EU. While one component of this regime manages residual waste, turning that to energy, another seeks to recover other parts of the waste stream for recycling and reuse. Within the EU, this is now regarded as the normative model of waste management, although it is acknowledged that there are potential tensions between the two strands, with there being evidence that high levels of energy from waste tend to correspond with lower levels of recycling.[24] Aside from these tensions, it also needs to be recognized that this is a regime that depends first and foremost on discard being classified as waste: for there to be secondary resource recovery there have to be wastes – and a lot of them.

What can be said of the world beyond the global North? This is a world that continues to depend heavily on landfill, although changes are occurring. To examine the broad contours of those changes, the focus shifts to consider first China and then India.

China's emerging waste regime has resonances with, but also important differences from, the EU's. At the heart of this is the notion of creating what is called an 'ecological civilization' resting on waste reduction, the efficient use of resources and using extracted materials in a circular manner. In China, therefore, there has been a twin-pronged emphasis over the past decade that brings together industrial ecology (Chapter 5) and waste management. Industrial policy has pushed the development of industrial ecology linkages, far more so than in the global North.[25] This has been seen as a means to reducing wastes from manufacturing, while there has been a simultaneous focus on reducing the amount of municipal waste being generated. So, in China, from the outset of the 2000s, what has been evolving is a waste regime in which waste has been seen as wasted materials, or a loss of resource – a situation that only began to take hold in the EU from 2015, with the turn to circular economy thinking and, later, directives.

The emphasis on waste reduction in China emerged in the mid-2000s, at the point when China exceeded the US in the amount of municipal waste being generated and landfilled. Attending to China's municipal waste problem, however, was (and remains) a matter of finding solutions to what is seen as a volumetric problem. Unlike the situation in Europe, landfill as a technology was not seen as a bête noir. So missing here is any sense of an attack on landfill – a situation that reflects the lack of sway in China of climate science in the policy domain of the time. Instead, landfill was recognized as offering a cheap, acceptable disposal technology that also allowed for the capture of methane gas. Its problems, such as they were seen to be, were more that it wasted extracted materials (that is, it didn't fit easily with the ideals of an ecological civilization) and that it failed to offer a solution to the 'too much waste' problem, since (at best) it merely compressed waste. Indeed, even in 2005, projections were suggesting that China would need an additional 1,400 landfills to deal with the volumes of waste anticipated by 2030.[26] Identifying landfill's problems in this way proved exceptionally favourable for the waste management technology that could reduce waste in volumetric terms (by ~90 per cent): incineration. What has followed in China is a programme of rapid growth in incineration capacity, as the 12th Five-Year Plan (2011–2015) created a set of conditions favourable to developing incineration at pace and at a scale suitable to serve the rapid urbanization rates of the major cities.[27] Preferential taxes, gate fees, feed-in tariffs and tax exemptions have all been deployed, with the goal of creating an environment attractive to big capital.[28] The result has been a 25 per cent annual growth rate in the number of plants with 331 incinerators handling

133 Mt per annum, or 44 per cent of municipal solid waste according to 2018 figures.[29] The Chinese municipal waste regime, then, appears very much a waste-to-energy regime – or is it?

What has emerged recently in China is what looks to be an attempt to overlay this waste-to-energy regime with something more like the municipal waste management practices found in the EU. In July 2019, Shanghai (a city of 26 million producing 9 Mt of municipal waste per annum) saw the introduction of a pilot programme that sought to 'educate citizens in the habit of waste sorting'.[30] Household responsibility for sorting and segregating wastes (or 'source segregation') had arrived in China – or, more accurately, been reintroduced.[31] To western eyes, the four categories look extremely familiar: recycling (effectively 'dry' recyclables), hazardous, wet (effectively food waste) and residual (everything else). However, it is a 'bring' system (where people take discard to bins located in a neighbourhood) rather than a household collection system in which bins are attached to dwellings. The whole was due to be rolled out to a further 46 cities in 2020, although the global COVID-19 pandemic intervened. This, however, is source segregation with Chinese characteristics, for, unlike the situation in Europe, it comes with strong financial and social imperatives with which to conform. There are fines (to the equivalent of roughly $30) for inappropriate categorization by individuals (much more for businesses), a surveillance system and knock-on effects for non-compliance, for fines are recorded and that record acts to restrict access to other widely seen to be desirable private goods, notably private education and international travel. The habit of source segregation, then, has been connected to patterns of consumption (Chapter 3). In such a way, I would argue, recycling habits are becoming increasingly entwined with the emerging idea of the Chinese citizen consumer. In a further twist, the resultant panic over 'doing the right thing' with household discard already has seen source segregation mediated digitally: an array of apps have been developed to assure people that they are sorting and putting their trash into the appropriate categories and bins, while in a rather different departure, 'waste collection services' have sprung up, advertising their services on Taobao, the online shopping platform.[32] There are two points I want to make in relation to these developments.

First, in waste management terms, an open question is just how effective such a strategy will be. Already there is strong evidence that societies with high levels of energy from waste (incineration) struggle to realize high recycling rates. So, there are questions over how much recyclate will be realized, even with a punitive system in place. But more than this, the aim appears to be to reduce the amount of (wet) waste heading to incineration: there are goals that 39 per cent of food waste will be captured by 2030 and 57 per cent by 2050[33], all of which will require the rapid expansion of an anaerobic digestion network. This is to seek to remove from the municipal waste stream the very fraction that is widely seen to have caused early Chinese energy-from-waste plants their biggest technical struggle.[34] At the same time, it is to remove a significant volume of waste feedstock, which, unless it is supplanted with other more calorific waste, will impinge on business economics. Down the line, China may find itself with excess incineration capacity.[35]

A second set of observations is grounded in how all this fits with the stated goal of creating an 'ecological civilization'. Tacit in the roll-out of western-style source segregation programmes is the erasure of informal sector recycling.[36] A crackdown on informal sector processing sites in key cities has led to their demolition and fuelled a process of geographical displacement.[37] At the same time, an evolving programme of environmental regulations has increasingly prohibited informal sector access to discard, while incentivizing big capital to introduce recycling initiatives. The roll-out of source segregation is but the latest in a long line of such measures. It seeks to bypass long-established cultural practices, whereby itinerant street traders purchased discard from Chinese households for small amounts of money and then sold them into the informal sector reprocessing system.[38] Instead, much as in the global North, households are now expected to hand over their discard for nothing, while doing the sorting work themselves and, unlike in the global North, face punitive sanctions for getting this wrong.

What this amounts to is nothing short of the erasure of the waste commons in China and its enclosure as private property. Discard is now something to which Chinese consumers no longer have rights; neither do informal sector traders and processors. Rather, in the name of an ecological civilization, the rights to discard have been appropriated by a coalition of interests that connects local government with capital, for sorted materials at city scale provide the volume of baseline feedstock that big recycling capital wants and needs if it is to turn a profit (Chapter 5). It cannot turn that profit if large amounts of value have already been extracted from the waste stream by the informal sector. In shape, if not in precise form, China has established a waste-to-resource regime much like that in the EU, although here there is little by way of environmental gloss. Rather, at every level, the regime choreographs discard to ensure maximum throughput for maximum profit. By definition, this is a regime that will require of its consumers to keep on consuming, and to legitimate that consumption by sorting their discard appropriately.

In India, at least outside of the megacities, what is emerging is pretty much the mirror opposite. Here, we find a waste reduction regime grounded in high levels of recycling and composting, with such residual waste as there is mostly continuing to head to landfill, aided and abetted by methane capture's qualification in developing countries for Clean Development Mechanism payments. So, here we see the very converse of the taxation regimes that characterize landfill in parts of the global North, with carbon accounting measures working to prop up landfill as a technology rather than making the technology increasingly uneconomic. Energy from waste in the form of incineration, by contrast, is as yet very much a minor part of Indian waste management. At the core of this new regime are waste collectors and pickers.

Back at the turn of the 2000s, things did not look as if they were going to be heading in this direction. Indeed, the direction of travel looked as if it would have considerable parallels with what has gone on in China. Research on the plastics recycling sector in Delhi, for example, identified that the beginnings of a rise in door-to-door municipal waste collection services, combined with an apparent turn to public–private partnerships (PPPs; albeit with non-governmental organization involvement), was closing off access to the waste commons for both waste collectors and waste pickers.[39] It suggested that 'corporate capital [would be] … interested in such a traditionally low status area of work because of its potential for yielding exorbitant profit'[40] and pointed to the potential for oligopolistic conditions to emerge. Contemporaneously, attempts to clamp down on the plastics recycling sector seemed to confirm a process of ring-fencing municipal waste for big capital, with recycling enterprises being handed out enforced closure notices on account of a judicial intervention that sought to eliminate 'polluting' and 'non-conforming' industries. So, informal sector recycling faced much the same challenge as in China: it failed to fit a notion of a modern India.[41]

By 2014, the situation had completely turned around. While the imperative to clean up remained, the identified pathway to that had changed. The means to this is Nahandra Modi's Swachh Bharat Mission, and the 2014–19 plan that accompanied this, which had the goal of achieving an open defecation-free India combined with 100 per cent scientific management of waste. Included within this are the 2016 Solid Waste Management Rules, which mandated municipalities to introduce source-segregated collections (of 'wet', dry and hazardous wastes) from households and businesses, but which stipulated that the arrangements municipalities put in place had to involve the informal sector, that is, waste collectors and pickers. What has emerged is a patchwork of different systems across different states, and different cities within the same states, but many of the examples of best practice feature highly decentralized systems that emphasize stakeholders and partnerships. Many have realized impressive results.

Often dubbed a 'waste-to-wealth' system, rather than a waste-to-resource system, the kernel to this success is the enrolment of the former informal sector in collection and sorting operations, often under cooperative enterprise arrangements whereby workers are rewarded with a share of the profits on sales of recyclables and compost. These door-to-door collection systems are highly labour intensive. Typically, hundreds of collectors service small neighbourhoods (of roughly 500–600 households) and they do so daily, or twice daily, using rickshaws as collection vehicles. Then other collectors and pickers work on the segregation and sorting operations that ensure that the maximum recyclable value is recovered from collections, while others work on neighbourhood-scale composting and biogas operations. A case in point is the widely celebrated instance of Ambikapur in Chhattisgarh, which is serviced by an 'army of Safai Didi (or cleaning sisters)'[42] numbering over 9,000. Chhattisgarh has the accolade of India's cleanest state. Ambikapur is the cleanest city in the cleanest state and sits as the model for a landfill-free small city – one that is trumpeted across developing countries.[43] Its 48 wards are divided into 17 modules of roughly 600 households, each serviced twice daily by waste collectors on e-rickshaws. The collectors deliver the waste to one of 17 solid and liquid waste management centres, where they are pre-treated prior to heading to either further sorting or composting. In the case of the recycling operations, a tertiary segregation centre utilizes 156 categories to aid sorting and segregation, with the results of this being offered for sale. A parallel strand of composting operations produces varying forms of compost for local markets, as well as biogas and animal feed.

While the Swachh Bharat Mission has had conspicuous success in India's small and medium-sized cities, the bigger challenge undoubtedly lies with its vast metro areas, most especially Mumbai, Delhi and Kolkata. The academic literature from the early 2000s spelled out the scale of the problem: in Mumbai, the then 13.8 million population was generating 8,000 tonnes of waste a day; in Delhi, 10 million were generating 6,000 tonnes per day; in Chennai, 5.8 million were generating 4,000 tonnes per day; while in Kolkata, 8 million were generating 3,000 tonnes daily.[44] At the same time, it foregrounded the unsustainability of a waste management system that was in practice entirely reliant upon dumping, highlighting an increasing lack of capacity along with unsanitary conditions. By the end of the 2010s, the situation had barely changed. The Ghazipur landfill in New Delhi, which opened in 1984, is now widely reported as 'higher than the Taj Mahal' and as 'a Mount Everest of trash that needs aircraft warning lights'[45], while descriptions of it, and its counterpart outside Mumbai (Deonar), echo the early 20th-century observations of J.C. Dawes on the South Hornchurch dump outside London (as mentioned earlier). Fires and slides of unstable waste generated by indiscriminate tipping abound, along with the odour from putrefying wastes and smoke. Small wonder that a 2021 review remarked: 'the dumps in Delhi have to be identified as the largest, least regulated and most hazardous in the world'.[46] The question is, what is to be done about this?

That question has no easy answers. It is, for example, far from guaranteed that a decentralized, highly labour-intensive system, which relies on turning large quantities of food waste to compost and biogas, can be scaled up to megacities of many millions, where many live in high-rise apartments and where the demand for compost is currently limited. With a turn to urban food production, that could change, but for the moment, the face of modern

21st-century urban India is much as elsewhere in the cities of the South – glitzy, sleek and most definitely not about getting up close and personal with compost. Add to that, there were multiple operating problems with the first generation of Indian big city composting plants, which meant that much of the output ended up having to be landfilled. This leaves scaled-up composting plants facing major technical questions.[47] A similar set of challenges have faced the only other technical solution for municipal waste: energy-from-waste incineration.

Unlike the situation in China, India's past experience with incineration has acted as a brake on the development of this sector. Time and time again in the academic literature, the case of Timarpur on the outskirts of New Delhi is recited.[48] This plant, which utilized technology from the global North supplied by the Danish firm Mijotechnik, opened in 1989. It was designed to handle 300 tonnes/day. So, it was a small plant by modern standards. However, the plant ran into serious operational difficulties. It was forced to shut down after only 21 days and was permanently closed after 6 months, with the cause widely identified as the low calorific value of the input waste, due to its high moisture content. Seemingly, this stands as a monument to failure that in turn indicates an Indian exceptionalism, namely that energy-from-waste incineration will never work on the subcontinent. This, however, is something of a myth, and it is countered by at least five plants that have recently come on stream and/or are in the process of build.[49] The material composition challenge is the same that faced, and that was overcome by, Chinese energy-from-waste facilities. That they were overcome is an indication of the degree of policy support for the energy-from-waste sector in China, something that remains conspicuously absent in India. But without the full package of feed-in tariffs, gate fees and preferential taxes, coupled with strong PPPs[50], it is unlikely that the energy-from-waste sector will engage fully in the Indian municipal waste market.

Elsewhere in the world, an emerging trend is for the big cities to bite the bullet with energy-from-waste solutions and to provide the package of conditions conducive to their set-up and operation. In recent years, Mexico City, Doha (Qatar) and other big cities in the Middle East, as well as Addis Ababa in Ethiopia, have all signed contracts for the development of energy-from-waste incineration facilities on the scale that characterizes the big Chinese cities. The Mexico City facility, for example, under contract to the French transnational Veolia, is set to handle 1.2 M tonnes per annum (and up to 13,000 tonnes/day), generating sufficient electricity to power the city's metro.[51] This is twice the size of any current facility in France and indicates the projected scale of waste generation in Mexico City over the next decades. In the face of estimates of 70 per cent and more of the world's population becoming urban in coming decades, with many living in megacities, this trend is significant. It highlights the intimate connections between systems of waste management and patterns of urbanization, suggesting that energy from waste is becoming the normative mode globally of managing municipal waste at the big city scale.

In my view, this is not just a matter of the apparent technical and business economics advantages of this technology for dealing with the waste generated by what is often termed 'planetary urbanism' – or, greater densities of people living in greater geographical concentrations of urban life. Yes, energy from waste offers cities the immense attraction of a volumetric reduction in waste, and yes, it is the most cost-effective solution for managing the wastes of dense populations, but its attractions go beyond this. Critical here is that through its capacity to produce electricity, the technology enables and sustains particular visions of cities. We see this particularly clearly with Mexico City, where the capacity to power a public transport system from the city's waste has won out over alternative, messier solutions to managing the city's wastes that would involve waste collectors and pickers and multiple waste-processing hubs. In other words, it is energy from waste's capacity to accommodate the modernist dream and to perform the ultimate alchemy, of turning messy, thoroughly material waste to clean, invisible electricity that is key to its current expansion, for with transit systems comes growing land value and enhanced potential for property-based development. Moreover, in generating power from waste, energy from waste allows the built form of the current megacity to go unchallenged, even in the face of the biggest challenge that many of these cities face, namely climate change. Electricity may power transit systems but it can also be turned to power air cooling systems, and it is surely not inconceivable that in the near future the technology will be hooked up to this end, marketed perhaps as the means to resilient living in cities located in increasingly hostile climatic environments. What this amounts to, I suggest, is the opening up of a huge market of the world's big cities to the energy-from-waste sector. In the next section, we turn to look more closely at this sector – its firms and corporate structure.

The energy-from-waste sector

There is a lot of 'hot air' around energy-from-waste facilities around the world, most of which is generated by the global anti-incineration lobby.[52] That lobby has more purchase in some parts of the world than in others, depending on the extent of the tradition in energy from waste and public trust in engineering science. However, its focus, typically, is on emissions, mostly dioxins.[53] Dioxin emissions act to catalyze public opposition and to mobilize campaigns, mostly in opposition to siting decisions and planning applications. If the plant is proposed in a developing country, waste pickers too are enrolled in anti-incineration campaigns, with their livelihoods being seen to be threatened by these plants. What all this does, I would suggest, is deflect attention away from what energy-from-waste plants actually are and what they do. It is to turn them into an emissions factory – or, it is to see waste as exclusively about pollution. That, I think, is unfortunate because it obscures what they really are, which is a complex power plant that has waste as a fuel. As I show in this section, it is only when we see energy-from-waste facilities as this that it is possible to see the array of big capital and financial interests that underpins this sector, and for which this sector is absolutely a major source of profit. It is useful to divide this into the technology and engineering firms that make the equipment, the waste management firms that operate it, and the finance and investment firms for whom the sector is of increasing interest for the return that it offers.

Fixed capital – on grates, boilers and proprietary technologies

Let's begin with the technology and the plant itself. At the heart of any modern energy-from-waste facility is a furnace grate/s, a combustion system and a steam boiler. Then there are the furnace feed systems, the turbines and generators and electrical equipment that generate electricity, the flue gas cleaning systems and a complex IT control system. All these systems interconnect and they need to be adjusted according to what is going on with the state of the fuel as it burns through, typically over the duration of an hour. So, these plants are highly complex, not just in the engineering that underpins their construction but also in operational terms. They are, correspondingly, hugely expensive to build. That has knock-on implications. Plants like this take several years (typically around a decade) to achieve financial payback on the original investment outlay. So, they are plants that are designed to have a long operational life. Typically that would be of the order of 20–30 years, with scheduled repair and maintenance being integral to their continued operation. In turn, that knocks on to the contracts that underpin operations, with the anchor contracts – provided by municipalities – being typically of the order of 20–30 years.[54] As buildings in the urban landscape, too, energy-from-waste plants are features that are designed to endure. Indeed, some have bespoke, award-winning architectural surrounds, which work to signal that these plants are urban monuments, contemporary cathedrals to our collective consumption, and the role of the city in this.[55]

Technology like this is far from universal. Much of it is proprietary, protected by large numbers of patents, with access to various technologies strictly controlled through multiple licensing agreements. In these circumstances it is unsurprising that a small number of firms have carved up a very large part of the growing global market in energy from waste. All of them are firms headquartered in the global North, and all of them operate internationally. Most have a long history in the power sector, with expertise in furnace and grate design, boiler construction and power generation that goes back at least to the beginnings of the 20th century and in some cases to the mid-19th century. These are firms that literally powered much of an earlier phase of capital expansion and urban development in Europe and North America, and they are now leading a further phase of capitalist development in which power generation has become the means to capturing more and more of the market in waste.

Take one of the lead firms in the design and manufacture of grate technologies, Martin Gmbh.[56] Martin is the leading global supplier of grate technologies for energy-from-waste facilities. It provides an excellent example of how proprietary technologies and licensing agreements, along with partnerships and acquisitions, combine to grow market share. Headquartered in Germany, Martin owes its current market position to a pioneering innovation back in the 1920s, which was the reverse-acting grate. The principle behind this grate is that fuel ignites more easily when a glowing mass is pushed back against it and under it. So, the reverse-acting grate is constructed in a stepped sloping arrangement, with the combination of gravity and reverse movement working to circulate burning fuel, thereby creating the conditions for optimal burn. The 1920s saw Martin sign a number of agreements with boiler manufacturers in Germany, which led to the grates being used primarily to provide combustion for the coal-fired industrial boilers of the time, many of which were used to generate electricity. It was not until the 1960s that a new market in the form of energy-from-waste plants began to emerge in Europe, with many of these plants being used to power district heating networks. That new market led to a signed agreement between Martin and the French firm CNIM in relation to the French market, and by the end of the 1970s, Martin grates were the dominant technology firing European energy-from-waste plants. Simultaneously, an agreement with Mitsubishi Heavy Industries in Japan allowed the technology to be adapted to deal with the (much wetter) municipal waste produced there and served to open up this market. More recently, in the 2000s, the acquisition of Alstom allowed Martin to secure two further grate technologies, including the SITY 2000 grate. Along with licensing agreements with Chongqing Sanfeng Covanta and China Everbright International (2013) this has opened up the Chinese market for Martin's grate systems. Currently, 31 per cent of Martin's combined total of global plants are either operational or under construction in China (182 of 588 plants worldwide). They have a capacity of 175 k tonnes/day and provide the combustion grates for over half the energy-from-waste plants in China. At the same time, CNIM Martin was opened in Chennai in 2017. The licensing agreement signed with China Everbright also extends to a few African countries – Ethiopia and Ghana at present. That is suggestive of positioning with respect to a yet-to-fully-emerge African market.[57] Meanwhile, the 2010s saw a further patented innovation – the reverse-acting grate Vario – developed for the European market. Offering enhanced operational flexibility and targeted agitation, this technology is a means to attending to the differential burn qualities of residual waste – of critical importance as the calorific value of residual wastes reduces in Europe in response to the extraction of more materials for resource recovery.

In a very real sense, then, it can be suggested that Martin's grates have captured a very large share of the current global market in residual waste, while licensing agreements signal intent in new and emerging markets. What lies behind this market domination, in part at least, is design flexibility. The grates are modular in design, with anything from one grate to eight parallel lines of grates being possible. In such a way, Martin grates can be designed for the differential needs of larger and smaller cities. At one end would be something like the Kitakami plant in Japan (2018 start-up) with one line and a plant capacity of 182 tonnes per day; at the other would be Shanghai Laogang II (2019 start-up) with eight lines and a plant capacity of 6,800 tonnes per day. But what also lies behind Martin's success is that this is a mature technology, complete with a long-standing set of partnerships and agreements, which work to guarantee relations across the supply chain and which are underpinned by service agreements. Cities end up purchasing these grates and these systems precisely because they offer what is regarded as a 'bankable technology' that is underpinned by often decades of collaboration between firms. That collaboration extends not just through the construction phase but also into the servicing and maintenance of operational facilities.

Another major firm in the sector, Babcock & Wilcox, provides a similar picture, but this is a firm that has its origins in boiler manufacture rather than in combustion grates. Founded in the US, its original patents go back to the mid-19th century; so, these boilers literally powered the electricity that was essential for the development of many industries in late 19th/early 20th-century North American cities.[58] Later in the 20th century, Babcock & Wilcox developed another boiler more suited to powering waste-to-energy plants, and the boilers started to be used in such plants in the US and Canada. However, it was not until the 2000s, through the acquisition of various business units of the Ansaldo Vølund Group and then the creation of Babcock & Wilcox/Vølund, that Babcock & Wilcox acquired grate technologies that could compete with the Martin grates and were able to fully enter the market as suppliers of energy-from-waste plants. Their history works to show, therefore, just how critical combustion grates (and their proprietary control) are to position in the sector.

Martin and Babcock & Wilcox/Vølund are prime examples of global North transnationals, for whom proprietary technologies, detailed knowledge (in the form of expertise in combustion and power generation, and systems control) and vertical integration, through supply chain partnerships, underpin business success. In that regard, the energy-from-waste sector is no different to say big pharma, aerospace or auto-manufacture. These arrangements ensure that the value in the world's residual waste is being captured by the global North. As more and more cities in the world face the challenge of reducing their waste, so more and more of the world's waste is open to capture through these technologies, precisely because it is incineration as energy from waste that is the means to addressing the scalar challenges of waste, through volumetric reduction.

To that, we can add that the energy-from-waste sector is not just dominated technologically by a small number of firms headquartered in the global North, but that it is a sector that is also increasingly capturing the construction of these technologies for big capital in the global North – in this case, engineering and construction firms. Building plants of this complexity is not easy. Often, early efforts at developing energy-from-waste capacity in developing countries ran into operational difficulties precisely because of shortfalls in appropriate local engineering skills. It is important to recognize that that shortfall is not unique to emerging markets. An article in *Construction Manager Magazine* in 2019, for example[59], remarking on the difficulties encountered in the energy-from-waste sector in England, pinpointed a lack in engineering skills there as a reason for the mounting difficulties being encountered. Likening the approach of English-based construction firms to thinking they were building 'kit in a shed' rather than 'a complex power plant', it indicates the problems that can occur if in-situ engineering skills are absent. So it is that we find the kind of situation that is occurring with respect to the latest waste-to-energy plant commissioned by Shenzhen in China, where the consortium, led by Babcock & Wilcox/Vølund and using their DynaGrate® system, is being assembled first in Ebsjerg in Denmark and then shipped to China in 48 parts.[60]

Operating capital – or, waste management as business

Fixed capital, in terms of the design and construction of plants, is the first of two key ways through which the energy-from-waste sector enables capital to turn waste to money. The other is through their operation. This relies on the plant feedstock, or residual waste. In the energy-from-waste sector, much as in many other areas of resource extraction, the rights to that feedstock are appropriated by capitalist firms. In that regard it is helpful to draw an analogy between an energy-from-waste plant and an oil rig, say, or a cobalt mine. Mines give firms often exclusive access to resources, through ownership, lease deals, concessions or extractive rights purchases. In the case of energy from waste, the rights to the operation of an energy-from-waste plant are the means to securing exclusive rights to the residual waste generated by particular cities, or parts of a city. It is the value in this residual waste that allows those who have the rights to this waste to turn it to money. How does this happen?

Much as with oil, copper, nickel or any other metal, we can think of the value in the residual waste of cities in terms of assets and revenue generation, that is, as both stocks of value (here, the estimated total yield of waste over a given period, typically 20-plus years) and the value realized by its processing or treatment in the plant. Let's take the second of these sources of value first.

Burning waste realizes energy in the form of heat, which is then used to power the boiler system in an energy-from-waste plant. That system can be used to generate heat, which can then be sold into local heating networks. The heat can also be used to power steam turbines that in turn generate electricity, which also qualifies for renewable payments. That electricity, too, can be sold to another user, be that the grid or a dedicated facility, typically one with a high demand for electricity. With any mined resource, the material quality of it has an economic effect, often determining the economic viability of a field as a mine. In the case of residual waste, the parallel to the purity of a particular mineral ore is the calorific value realized by burning the residual waste. This is variable, with geography and seasonality both exerting effects. As we have already seen, the residual wastes generated in China and South Asia, in general, yield lower calorific value than the residual waste that was being generated in Europe and North America, particularly in the latter decades of the 20th century.[61] A broad rule of thumb, however, underpinned by basic physics, is that the higher the calorific value realized by combusting the waste, the more the heat energy that will be produced – and the more money it therefore realizes. This is why calorific value matters. It is not just that the combustion of certain waste yields lower calorific value, or that some wastes are harder (and therefore more expensive) to combust; it's that its combustion yields less heat, so there is less value to be extracted from this waste.[62] That means that more waste will need to be burned in order to realize an equivalent amount of energy. A corollary is that the stocks of wastes of some cities are worth more intrinsically than others. This can be anticipated to have an effect on the price municipalities are charged for their processing.

Turning to residual waste as a stock of value – this is where residual waste differs from conventional primary resources. There, stocks are finite. Once a reserve is depleted, or becomes too difficult and costly to mine, it is generally ripe for closure. The history of mining activities is replete with examples of such practices. As firms divest themselves of these assets, mines close down and leave behind communities that were largely dependent on extractive activities, often with devastating economic consequences. Residual waste, however, is not like this. Instead, it's more like water. It's one of those resources that are continually being replenished. With water that occurs through natural environmental processes; with residual waste, it's the process of human consumption that is the means to resource replenishment (Chapter 3). What this means is that, so long as human consumption carries on in the same shape and form, so residual waste will keep being generated. And in that statement is the nub of the contemporary dilemma over energy from waste, for while they might work to reduce waste in volumetric terms, energy-from-waste plants as economic entities require that human consumption keeps on feeding them not just sufficient waste to burn, but sufficient waste of sufficient calorific value for the plants to turn a sufficient profit. That is bound into the contracts that underpin them, with municipalities contractually obliged to provide a guaranteed volume of residual waste to certain material parameters, and the off-take electricity contracts that operators negotiate resting on that input. As the sector expands globally, then, so the demand for residual waste will only increase. This situation, of course, is very much in the interests of the corporations who operate energy-from-waste plants – the waste management businesses. So, having looked already at the corporate structure behind the technology that underpins these plants, let's shift to consider the rather different set of corporate interests that operate them.

I would preface my arguments here by stating that there is nothing intrinsic to an energy-from-waste plant that means they have to be operated by private capital. In practice, however, they mostly are. This is an effect of the expertise and skills they require to operate and the near ubiquity of the PPP finance model in their financing. That situation, in turn, has made possible the emergence of what are best described as global environmental service companies. These companies specialize in energy from waste, often as part of a wider portfolio of interests in the waste management sector. Typically, such companies assemble consortium bids in both the municipal and commercial markets. Their partners would include the kind of technology firms already discussed, whose role in this type of arrangement is to act as suppliers. It is the environmental service companies, however, who are seeking to capture the value in the residual waste itself. The market here is more fluid than that in relation to fixed capital provision but a good example of such a firm, and one with a strong global presence, is the French transnational Veolia.

Veolia's corporate history traces back to the mid-19th century, as a municipal water company treating urban water. Its beginnings, therefore, are in the urban sanitation movement of the 19th century (see previous section: Waste, Public Health and Sanitary Disposal) and the modernization of French cities – the most notable example of which is Haussman's redesign of Paris.[63] Veolia's first business interests were in the separation of drinking (or potable) water from waste water and their different treatments, and its first contracts were for the municipal water of Lyons, Nantes, Nice and Paris.[64] Immediately, then, we see a set of commercial connections between waste and water, which continue to this day and which go well beyond this corporation. In the 20th century, the company expanded into multiple municipal services, adding waste management, energy and metropolitan transport services to their portfolio. It also expanded internationally. Major restructurings in the early 2000s led to the divestment of the transport business and US water subsidiaries, and a refocus on core business: water, waste and energy. Those businesses remain core for the group into the 2020s. Veolia's waste services portfolio spans the full spectrum of waste management, encompassing hazardous/liquid waste management, energy from waste, commercial and industrial collection services, municipal collection services, and street cleaning, recycling and landfill. In terms of revenue generated, the most valuable of these in percentage terms are hazardous waste followed closely by energy from waste.[65]

In 2020, Veolia operated 685 waste-processing facilities globally, had nearly 179,000 employees, supplied drinking water to 5 million and wastewater services to 62 million, and treated 47 million tonnes of waste. With an annual revenue in excess of €27 billion[66] and listed on the Paris stock exchange, it is a global transnational comparable to a small to medium-sized global car manufacturer. But evidently, the ambition is to grow much further. An aggressive takeover of its main French rival, Suez, in 2020/21 was underscored by a clearly stated goal: to be, in the words of the company's chair and CEO, Antoine Frérot, 'the undisputed world champion of the ecological transformation'.[67] The ecological transformation in question is the quest to be at the forefront of greening capitalist economies through the development of sustainable, circular urban solutions in the water, waste and energy sectors. Through the takeover and subsequent restructuring of Suez as New Suez, Veolia and New Suez are seeking to grow market share in water, waste and energy services in new and emerging markets, as well as to consolidate their already very strong position in the European market.[68] The new combined entity is without question the major European firm in the sector.

One of the best illustrations of the effects of this domination is the UK.[69] In 2020, Tolvik Consulting[70] recorded that 54 per cent of the UK's input to energy-from-waste facilities was handled by just three firms: Viridor, Veolia and Suez. The acquisition of Suez by Veolia took that share of 2020 input to 32 per cent – or, nearly a third of the UK's energy-from-waste market waste had been captured by this one transnational corporation.[71] The remainder, however, was shared by a small number of much smaller players.

Patterns like this provide a ripe terrain for acquisitions, and so this has proved. Take the example of Urbaser, Spain's leading energy-from-waste supplier and one of the smaller players in the UK energy-from-waste market. Before the merger of Veolia and Suez, its parent company Urbaser SAU was the third largest environmental service provider in Europe. However, in 2016 it announced that it was getting out of the energy-from-waste sector.

Urbaser was acquired by Firion Investments – a Chinese investment firm directly controlled by a green industry fund (Huaya). Acquisitions like this are the means to acquiring technological and operational know-how. They also connect to the wider geopolitics of China's 'Belt and Road' infrastructural programme, in which gaining control of European infrastructural assets is emerging as a key means to reorient capital flows for Chinese benefit. But, as the Urbaser case also shows, the combination of this type of corporate landscape with key infrastructural assets is an attractive terrain for finance capital. I turn finally in this chapter, therefore, to consider how finance capital interweaves with, and increasingly drives, the global waste-to-resource regime.

Finance capital – or, how investment turns waste to wealth

Recent research, much of it conducted in the wake of the global financial crisis (2007–9), has done much to address the previous opacity of an increasingly financialized capitalism. Included here, alongside a welter of work on financial traders and trading, is research on the private equity sector.[72] I want to begin here because this is where finance capital intersects most readily with the corporate market in energy from waste.

Private equity firms seek to buy out small to medium-sized firms using high amounts of debt-based finance in which the assets of the buy-out are used as collateral. They look to improve these companies through the introduction of operational improvements and management efficiencies, prior to their resale in a short time window – typically less than five years. But what often happens with these 'leveraged buy-outs' is that, in the quest to extract more of a surplus for investors, their assets are used as a means to mortgage further loans. So, companies purchased by private equity firms can often find themselves not only having to service their original debt, but loaded with more debt repayments, while simultaneously paying not insignificant management fees to the private equity firm. The result is that healthy companies can end up in serious financial difficulties and needing to sell assets to manage their debt. For private equity firms and the investors in their funds, this is a pretty much sure-fire way to extracting surplus financial value out of the company. The fees alone (set at a multiplier of turnover) provide a substantial return to investors (comprised of wealthy individuals and institutional investors such as pension funds), and then there are the proceeds from any sales. Small wonder that this type of wealth-generating activity has been held up as exemplifying the parasitical tendencies of finance capital and as emblematic of 'casino capitalism'.

Small to medium-sized firms operating in the energy-from-waste sector are particularly attractive to private equity firms. They are so because these firms are asset heavy, so they offer plenty of scope for remortgaging, while the scale of their operations provide serious scope for management fee payments. Add to that – especially at the outset of the transition from the collection-disposal regime to the waste-to-resource regime – the array of firms seeking to establish and consolidate their position in what was then an emerging market. Not only did this provide a ready supply of firms ripe for private equity acquisition, it also suggested a strong market for their resale, with many firms looking to consolidate market position. Again, the UK provides a useful illustration.

One of the first examples of this type of activity in the UK occurred with the acquisition of Shanks' landfill assets by the private equity firm Terra Firma, for £531 million in 2004. Terra Firma sold those just two years later, along with assets of WRG, to the Spanish firm FCC for an estimated £1.4 billion.[73] Then there is the purchase of the UK waste management firm Biffa for £1.7 billion in 2008 by a private equity partnership of Montagu, Global Infrastructural Partners (GIP) and HBOS Fund Investments.[74] In 2010, under Montagu and GIP, Biffa acquired another UK waste management company, Greenstar – loading a debt package of a further £1 million onto the company's books. By 2012, with the effects of the financial crisis in full swing, Biffa was having difficulty servicing its debt and a 'fire sale' was being scoped out, with Montagu and GIP in negotiations with Veolia, Suez and Pennon Group (the parent company of Viridor). The negotiations with the big three in the UK market, however, came to nothing.[75] By 2013, Biffa had been recapitalized through a debt for equity swap involving an array of fund managers specializing in distressed debt.[76]

Private equity interest in the UK energy-from-waste sector remains strong. Some ten years on from the Biffa-leveraged buy-out and its aftermath, in early 2020 Pennon Group offered Viridor for sale – one of the major firms in the UK waste sector. This too has been purchased by private equity interests – in this case the US firm KKR, in a £4.2 billion deal.[77] Less than a year on, certain assets of it are already being sold: in 2021, KKR instructed the company to sell the vast majority of its UK recycling infrastructure.[78] The purchaser, reportedly, is Biffa, at £126 million. Beyond the UK, it is the same story elsewhere in the sector, with Covanta – the US's major energy-from-waste supplier – being purchased by the Swedish fund EQT for $5.3 billion in 2021.[79] While in Korea, private equity firm Affirm Capital is estimated to have made over eight times the purchase price in just four years on its sale of Korea's biggest waste management firm EMC Holdings.[80] No longer confining its interests to small to medium players in the sector, private equity is moving in on some of its larger companies.

Tracking private equity acquisitions indicates the extent of finance capital's interest in the global energy-from-waste sector. But there are further ways in which the two are interlinked. Beyond hedging by the major energy-from-waste firms, the most significant of these is through the financialization of residual waste – or, its assetization. Here, individual energy-from-waste projects – their plants, the present and future waste that they treat, and the population, households and businesses that generate this – are turned to assets. How does this work?

The favoured model for financing energy-from-waste projects, as we have already seen, is the PPP. This not only allows for private equity interests and banks to come on board in a tender bid; the special purchase vehicle (SPV) that is set up by the successful consortium also allows for residual waste to be turned into a financial asset, thereby opening the project up to an array of other potential investors beyond the original investors. SPVs are set up by all the major energy-from-waste operators at the point of contractual close as the means to ring-fence the financial risk on any one project. Since they are off the balance sheet, they protect the company from being contaminated financially by a project that runs into difficulties. So, they make sense in corporate risk management terms. At the same time, SPVs work to obscure the debt liabilities of companies, as was exposed in the case of Enron.[81] So, they are seen to be of benefit in accounting terms – basically because they protect the bottom line and stock price from any debt liabilities. What SPVs also allow, however, is for the holder of the SPV (typically the operating firm in an energy-from-waste consortium) to repackage some, or all, of the debt incurred on the loans, as a collateralized debt obligation, in which the asset (here an energy-from-waste plant) acts as the collateral against default. Collateralized debt obligations are derivatives. They are financial products that are a means by which firms free up some of the capital locked up in loan repayments, by selling that debt into the financial markets as bonds. In this way, we can see that an energy-from-waste contract generates two revenue streams that are turned to assets. One is the yield that comes from treating residual waste. This is a combination of the plant, the input waste and resultant energy resources that are sold as electricity – and it is this that the original investors are tapping into. The other is from debt refinancing – and it is this that is being sold (as bonds) into the financial markets via the SPV.

As with many other forms of urban infrastructure, then, residual waste has been turned into an investment opportunity, through financial techniques of securitization. The most familiar examples of such practices involve municipal water. Here, guaranteed revenues from households (in the form of water bills) have been shown to underpin their translation into securities, bonds and equities suited to investors.[82] It is a very similar set-up with residual waste: guaranteed revenues flow from households through the medium of local council taxes, which include payments for waste services, and the annual payment charge made by the municipality to the supplier/s. Small wonder, then, that many of the companies that have got into managing and operating urban waste infrastructure are the very same companies that have an extensive stake in urban water. It is the same model of business.

What of investors? If one turns to the literature aimed at investors and fund managers, it is not difficult to see why energy from waste is currently an attractive investment opportunity. In July 2021, Fitch Ratings, for example, were highlighting anticipated returns in the 8–14 per cent range and they spotlighted the strong performance on yield, growth and safety of some of the companies already discussed.[83] They summarize the sector as offering stable, recurrent earnings underpinned by fixed power tariffs, predictable volumes of household waste and long-term concession treatments. Income from contracts is seen to be backed by cost-indexed fees and minimum volume guarantees, and the differential earnings from big cities versus smaller scale facilities are emphasized. With anticipated growth in waste consequent upon rising levels of global consumption, this is low risk for medium return terrain, hence why residual waste has become a portfolio banker for investors, from wealthy individuals to fund managers to pension funds.

More generally, we see in such statements how the PPP model as applied to residual waste benefits finance capital and how waste begets wealth. Just as significantly, it should be noted that there is nothing here to promote the reduction of waste. Rather, it is the converse. Such is the appetite for energy-from-waste projects from finance capital that it's more, not less, waste that is required to service its insatiable demand to extract a surplus, hence the burgeoning growth in the number of energy-from-waste plants globally. That this is masked by the rhetoric of waste reduction that surrounds energy-from-waste technologies should be no surprise, given the widespread turn to technocratic forms of politics. But such rhetoric needs to be seen for what it is: namely a fallacy that depends on diversion from landfill to state its case. It's a reduction in volume only. The reality, instead, is that with residual waste, as in so many other spheres of the real economy, and to use a phrase, 'Wall Street has captured Main Street'. The waste-to-resource regime that is emerging globally is a regime that is not only capturing the rights to the world's wastes in the hands of a small number of transnational corporations, but – through the ubiquity of the PPP model and the SPV – it is also simultaneously being configured to suit the interests of finance capital. In the ultimate of ironies, in this new waste-to-resource regime, waste – the textbook example of the economic externality – is no longer the stuff of disposal. It has not only been commodified but it has also been assetized. As a result, the waste of the world goes beyond the terrain of 'muck and brass'. It is now inextricably linked with wealth generation and accumulation.

7

Future Directions

Or, rewiring waste through the three Ds (decarbonization, digital and discard)

Having begun this book with my motivations for writing it, I end it by underlining its messages or implications for policymakers and for each and every one of us as consumers. At one level they can be stated simply and starkly: we cannot solve what is called the 'global waste problem' by seeking to eliminate or get rid of waste. Neither will we make significant inroads into the problem by focusing on the stuff of waste, or discrete categories of waste: food waste, packaging waste, e-waste and the like. This is because waste is not just material, physical stuff. It is more fundamental than that. It is a category of value and a way of ordering lives. We will never, therefore, get rid of waste; rather, it's a category that will continue to be made and remade in the course of everyday life.

It is also important to acknowledge that the category of waste also includes stuff of negative value – the kind of stuff that is harmful to living organisms. I have said less about this kind of stuff in this book, but the examples are legion: radioactive waste, asbestos, PCBs, clinical waste, for example. This is important to acknowledge because these wastes require specific technical waste management solutions to deal with them safely, and at least one of those wastes currently does not have such a solution. However much it might wish otherwise, humanity cannot do away with these types of wastes. Some, like radioactive waste and asbestos, are in the 'forever' category; they are a legacy of previous generations' approach to the material world and a potent reminder that what is made never goes away but only transforms state, that is, it continues to exist in material form in some way or other. Others, notably clinical waste, are an inevitable effect of the ongoing need to care safely for life, but the same laws apply.

We have to proceed, therefore, by recognizing that there will always be waste and that certain wastes will always need to be managed appropriately. Where we have a choice, however, is in what else enters this category we call waste, and in what volume. There is no inevitability with regard to what of consumer discard becomes municipal waste. As I have shown in this book, its material characteristics, what is placed in the category 'municipal waste' and the volume of material that becomes that category are all the result of societal, political, economic and cultural choices.

Currently, the choice that the countries of the global North and China, and a growing number of middle-income countries, are making is to turn more and more consumer discard to waste, with increasing intensity, while maintaining that we are doing exactly the opposite and reducing waste. That fiction is supported by a focus on technical and political definitions of what counts as waste. But, as I have shown, capitalizing and financializing consumer discard will result in heightened demand for more stuff to be declared waste. Moreover, the financialization of waste that underpins the global shift to incineration is being compounded by enhanced materials recovery and recycling and the increasingly common regulatory demand for products to include greater recycled content. Such demand rests on continued primary resource extraction, for few goods are manufactured solely from recycled content. Inevitably, too, demand for more recovered materials relies on enhanced demand for consumer discard. Eking stuff out by enhancing its efficiency might sound attractive. It appears to stretch finite resources further and to allay concerns about resource scarcity. But what increased demand for recyclable materials actually requires is that (more) consumers consume, and then discard more and more stuff to become feedstock to supply a burgeoning recyclate recovery industry. Far from attending to the global waste problem, the choices that have been made in relation to waste the world over will demand more material consumption of the world's consumers, and hence more production, while ensuring that more stuff becomes waste. This is a capitalist fix to the global waste problem.

So, how might things be otherwise? How might we get off this wheel? In this last chapter, the focus is not on identifying definitive prescriptions or a 'do this' list of recommendations. Neither do I make suggestions regarding the stuff of waste. Rather, and in line with the arguments made in the Preface, my focus is on articulating the implications of the arguments made in this book about the relationship between consumption, economies and waste. I pose these implications as a set of three interconnected challenges, each of which is related to the premises of this book, as laid out in Chapter 1. Each challenge starts from what needs to change if the connections between consumer-generated discard and waste are to be severed or, at the very least, weakened. In each case, I then make a series of suggestions as to areas, or sites, of potential intervention. The suggestions are aimed at policymakers primarily, in that they focus on where attention might usefully be directed to have the effect of reducing waste, but they will also be of interest to consumers motivated by systemic level change. In so doing, I hope to begin, rather than close down, conversation and to provide new openings and new possibilities for how we might begin to think about and address the global waste problem in more progressive ways, and in ways that link global waste more strongly to a future hinged to the decarbonization agenda and to an increasingly digitally mediated economy.

CHALLENGE 1

If waste is an effect of contemporary patterns and practices of consumption, how might these patterns and practices be reconfigured such that less discard finds its way into the waste stream?

Much faith has been placed by many countries in recent years in appealing to the restoration of erstwhile cultural values and practices to prevent waste. Notable here is the return to celebrate the virtues of thrift and frugality as waste-saving and waste prevention practices, respectively. I will say now that I think it unlikely that such appeals will have the desired effect – at least at scale. History shows us that thrift and frugality have only worked at scale in societies with command and control economies, that is, in conditions of war, such as occurred with rationing in World War II, or the planned economies of the Soviet Union and China. In market-based economies and societies, while there are undoubtedly some people who will willingly embrace these practices, mostly these are practices born of necessity. They are ways of life and ways of living with things and stuff that are closely bound up with living in conditions of poverty and austerity. For those whose lives are rather more comfortable, going without in generic conditions of plenty, or eking things out, become ethical choices, not necessary measures. Ethical choices are always hard choices. Mostly, though, they are understood to involve decisions taken by and affecting an individual. That is an effect of the rule of law and the connection of rights to personhood, and hence to individuals. That way of thinking has travelled through to frame how we think about 'the consumer' – as an individual. Indeed, we talk about and use the category of 'the ethical consumer' to signal a consumer motivated to care about consumption's consequences, be those felt by the people who make goods (the Fair Trade movement is the best example of this) or the environment (for example, vegetarians and vegans). Care in both these instances is seen to be expressed in the acts of purchase that are made (or not made) by a consumer. Incorporating thrift and frugality under the umbrella of the ethical consumer is but another example of caring about consumption's consequences. It signals a consumer who cares about consumption's connection to waste, and for whom the inevitable consequence of that care is a limit on purchasing. So, if the world had more ethical consumers who cared about waste, we'd be on our way to sorting out the waste problem. Unfortunately, I don't think it's that simple. Here's why.

The challenge that thinking about consumption as a meta-practice poses to ethical consumption is that it shows that a very great deal of consumption is not about an individual. Rather, it shows that it's social and it's relational, because when we engage in all the activities that together go to make up consumption (eating, drinking, cooking, playing sports ...), we are mostly doing social relationships with and through stuff. We can never get away from this because we live stuff-rich/stuff-dense lives. The challenge that faces invocations to thrift and frugality, then, is that to consume in this way requires us to deny, or at the very least put limits on, doing sociality through this thing- and stuff-rich world. More than that, living thriftily and frugally require us to do that to those we often hold most close, that is, those we live with and among. They also require us to do this when others all around continue to carry on as normal in how they live with and consume things. In other words, thrift and frugality require the altruistic sacrifice of kin groups – or, that related groups of people accept the moral virtues of going without, or going with less. That's OK when everyone is making the same sacrifice but it most certainly is not all right when others clearly do not. So this kind of sacrifice is dividing and it becomes potentially socially highly divisive. It will also have effects, not just on socially lived experiences but on the dynamics of our closest social relations, and it is particularly difficult for children and young people to navigate because having a parent who insists on going without and making do makes one stand out among one's peers not just as 'different' but as visibly different in things and experiences. At best, I would suggest, most of us who have a choice about thrift and frugality, most of the time, would find ourselves intermittent in our commitment to such ethical prescriptions, wary of their potential social effects and reluctant to impose our ethics on the lives of others.

So, if we're to pursue waste saving and waste prevention as consumer practices, I think we have to do so through more indirect means than through a direct appeal to values. One of the positive aspects of the COVID-19 pandemic is that it has shown how that might happen, through slowing lives down. One of the most immediately obvious effects of lockdown lives in 2020 – when the geographical mobility of lives across much of the planet was literally stilled – was that they generated less waste. Make no mistake, I am not arguing for such 'super-stilled' modes of life. But, the turn to flexible, hybrid models of professional working set in train by the pandemic, in which working from home became thoroughly normalized, stills lives sufficiently to have huge possibilities for both reduced consumption and waste prevention. A reduced carbon footprint from daily travel, less on-the-go food and drink consumption, less need for professional clothing and a reduction in the amount of clothing going through laundering cycles are just some of the effects here. However, it is the way in which decentred work intersects with and disrupts convenience that is perhaps its most significant effect in terms of waste prevention.

Provided firm constraints are placed on expanding the hours of work, remote, digitally mediated homeworking reduces many of the scheduling and compression constraints that result from different members of a household needing to be in geographically different places simultaneously. As a consequence, it allows for other activities to go on in the time that would otherwise have been spent commuting – and since most people pre-pandemic spent 1–2 hours per day doing that, that's a lot more time suddenly available. Most notable here is the time made newly available for food preparation and cooking – a trend that emerged immediately during lockdown periods, with readily discernible effects in the amount of food waste that was being generated. More people started engaging with cooking from scratch with basic ingredients; home baking became a craze – and none of this occurred with any appeals to thrift or frugality in sight. Rather, it was people with more time available finding pleasure and enjoyment in making their own food and meals, and people recognizing stock shortages and resorting to making do and eking things out without being told to. What that kind of shift also does is to bring into question the prominence given to convenience in the things that figure in contemporary lives. Less frenetic lives lead to a less intense sense of time as a lived experience. The need for, and appeal of, time-saving – be that stuff such as ready-made food, or meals made for you, or of devices that seemingly save you time – can therefore dwindle in these circumstances. They open the door to questions such as 'do I really need this?' 'Do I need to buy this kind of food any more?' Similarly, lives orchestrated in less frenetic ways can allow more scope for repair. They make it more possible to do the work of repair itself, but equally the being without of a particular functionality that is a feature of repair is potentially less awkward to accommodate in slower lives, making the discard-replace/substitute option less necessary. So, changing how and where many of us work is quite possibly the single most important thing that we can do to reconfigure consumption and its effects in discard. And the beauty of this is that it doesn't force people to reduce or go without; it allows them to realize this in and through the way they live their lives, and then to act accordingly.

The 'but' of course is that not all work can be reconfigured in this way. Not all work can be done from home – again, as the pandemic has shown. Healthcare workers, social care workers, waste workers, manufacturing and construction workers, retail workers, transport workers, supply chain workers, and trades persons are all examples of people whose jobs require, and will continue to require, movement between home and work, and in some cases to work in ways that require one to be constantly on the move. So, there is a very real prospect here of a twin-track divide opening up, in which one set of (largely professional) workers are able to reconfigure their lives in ways that generate less waste, while others, through no fault of their own, are forced to continue living in ways that put a premium on convenience and that therefore ensure they are generating more and more discard. Those twin tracks come together most clearly in the case of 'final mile' delivery workers. One of the ironies of less frenetic lives for many is that it has come with accelerated demand for home delivery and thus enhanced freneticism for those who supply goods to home-based workers. Enhanced freneticism places more and more reliance on convenience. A key challenge here will be to try to alleviate these effects. How might this be done?

One possibility here would be to scale up meals made from surplus food. At scale, there are three options for dealing with the problem of food waste: one is to collect it and turn it into biogas; a second is to feed it to bioprocessors (livestock); the third is to go big on the challenge of working with surplus food. The latter is where I think attention needs to focus. Rather than provide incentives that capitalize turning food to waste, there is a need to keep what has been produced as food in an edible state for long enough that it can be eaten safely. In conditions where food poverty is growing, not shrinking, where food poverty is a manifestation of widening societal inequalities, and where the climate emergency means that food shortages are more, not less, likely, I would argue that turning surplus food back to edible meals is an urgent political priority. Examples of 'pop-up' or one-off initiatives, whereby (often celebrity) chefs have cooked with restaurant leftovers, give an inkling of the possibilities here, but these sorts of one-offs need to be scaled up, and ideally in ways that use surplus food to attend to the problem that is food poverty rather than providing ethicality on a plate for fine diners. History gives us some templates, notably from those periods when there were food shortages and where there was a political necessity to attend to nations' diets just to keep people alive. In the UK, that model is provided by the 'British Restaurants' phenomenon – a programme of community-based, local authority-financed kitchens and basic restaurants that was set up in World War II, which ensured that a reasonable meal was available to all who needed it, at very low cost.[1] Adapted to the 21st century, such a model could be used not only to provide good-quality, nutritious meals at cost price for all who need that; it would provide a counter offer to the enforced reliance of so many 'key workers' on the privatized, market-based solution, which is fast food, on the go – the effects of which in waste were discussed in Chapter 2.

CHALLENGE 2

If current levels of waste generation are an effect of how consumer demand and 'the consumer' figure in mainstream thinking about economic growth, then how might consumer demand and the meta-category of 'the consumer' be reimagined to generate less waste?

The answer here is not the current hall of mirrors one that breaks the link between waste and growth by manipulating what counts as waste. Rather, if we are going to tackle that link in ways that really reduce waste, then there is a need to start with consumer demand. Specifically, there is a need to address what exactly consumers demand, and how much they demand of certain goods and services. And there is a need to do all of that in the context of the climate emergency.

My starting point here is with the nature of consumer demand. Given the connection of consumer demand to growth, it is very easy to argue that the only answer to the problem of consumption and waste is to think hard about what drives them both: the engine that is growth and the perceived political necessity for growth. When framed in this way, all too easily this becomes the kind of anti-capitalist/anti-growth argument that prevails in environmental circles. It is one that opponents reduce quite easily to the argument for a retreat to an imagined stone age. More modified positions recognize the political unpalatability of these arguments, their economic impossibility and their potential societal consequences. They ask instead how the levels of prosperity that come with a certain level of growth might be maintained without the dire deleterious environmental consequences that have accompanied economic growth in the period since the beginnings of industrial capitalism – the combined double whammy that is the global waste problem and GHG emissions. So, they acknowledge the wider societal benefits that come from strong and prosperous economies; they flag the political impossibility of anti-growth arguments with voters accustomed to living comfortable lives; and they recognize consumption's benefits – specifically, the pleasure and comfort it has brought to the lives of many people in the global North, and that it is beginning to do in other parts of the world. Consumption has allowed for living longer, mostly healthier and experientially far richer lives than was the case even for people living just two or three generations ago, whose lives were shorter and often defined by unrelentingly hard, mostly physical work. So this kind of thinking recognizes the rights of those living beyond the global North to have access to these sorts of lives. The problem, however, is that the world just doesn't have the resource capacity for everybody living on the planet to live the life of a global consumer – at least as this has been defined so far. Global consumers are defined not just by the physical footprint of the goods they consume, but by the reach of their geographical footprint, which is becoming simultaneously more extensive and intensive. The carbon consequences of this cannot be accommodated. How might this be tackled?

One option here is to go back to just what exactly it is that constitutes the category of consumer demand (goods and services), and to think about the balance of that in terms of resource consumption and carbon emissions. If we go back to Keynes' original formulation, stimulating consumer demand then was operationalized in the context of economies that were heavily oriented to the production of manufactured goods. Pulling policy levers to stimulate consumer demand at this time required that consumers buy manufactured stuff, and then lots more stuff, and that they replace this stuff. Inevitably, this kind of consumer demand had major implications for GHG emissions and for resource extraction. However, 21st-century economies, particularly those in the global North, are not in the same place. For a start, they are less reliant on manufacturing industry for growth; then, consumption itself has altered, with demand for consumer services (hospitality, leisure activities, entertainment, tourism) a more significant part of many consumers' budgets than the acquisition of physical goods. In part, that reflects the drive of globalization to ever cheaper manufactured goods, and the relativities of earnings in the global North to the price of those goods, but certainly the trend has been for more consumers in the global North to spend more money on eating out, travel and organized leisure activities than they have ever done previously. As the turn to 'the experience economy' ramps up, this will only increase. We can see this in the way that gifts have become less *something* and more *doing* something together – 'memory making' as it is often called, captured on a mobile phone and then constituted into the digital record of our lives via whichever social media platform takes our fancy. So, there is a sense in which consumer demand is potentially less reliant on physical goods than has previously been the case, even though it should be acknowledged that often this doing something entails the simultaneous consumption of stuff.

There is nothing about this trajectory, then, that says automatically that it will be less resource intensive than previously, but any world that is less stuff dependent certainly has these possibilities; and if that transpired it would inevitably result in less waste generation. This is because reduced demand for stuff slows the cycle of planned obsolescence. At the same time, once consumers see consumption in more experiential ways, it opens the door to the mainstreaming of different models of accessing physical goods, notably leasing, or rent-based models, rather than the outright acquisition/ownership model. We can see this already in the growth in the car rental market and, from a much lower base, in leased clothing. Goods leased, used and then returned are no longer going to become consumer-generated waste; rather, the responsibility for managing their discard will be with producers and retailers. Boosting the sharing economy is one way in which policymakers can drive producer and retailer responsibility further, while simultaneously incorporating more consumer goods within sustainable design and manufacturing remits.

That said, there are two key areas of this emerging new world of consumer demand that are, and will remain, intensely problematic for waste generation, and emissions generation more widely. They are: 1) the demand it places on digital technology; and 2) what we might term the 'emissions burden of experience'. Let's take each of these in turn, and again, I will suggest various possibilities for intervention along the way.

One of the most potent indicators of the increased dependence on digital technologies is the huge explosion in recent years in electronic waste, in the form of discarded digital devices. The mountains of this stuff that are now being produced globally belie any pretence to the lightness of digitally mediated life compared with its analogue counterpart. While we might carry around, and work and play with, less physical stuff, the reality is that the hardware that supports this rapidly declines in value, as we see with routine cycles of IT equipment replacement and upgrading, which occur in multiple workplaces. Software upgrades only intensify the problem. The same problem of planned obsolescence affects all digitally connected homes and consumers (Chapter 3). If the challenge of discarded digital hardware is to be attended to in any meaningful way, it requires breaking this intensifying cycle of planned obsolescence.

I see little prospect for policy interventions here that target the consumer. Rather, the most effective interventions will be on the supply side, but they will require coordinated regulatory action in relation to the tech giants and their tier one suppliers. Ultimately, proprietary software, rather than hardware failure, is what lies behind a great deal of electronic discard. So, unless governments enforce software support protocols on the tech giants – requiring that operating systems are to remain supported for x years, as a condition of sale in a given market – then it is unlikely that any dent will be made in this challenge. At the same time, and recognizing that hardware does require replacement eventually, there is scope to insist on minimum recovered material content stipulations with respect to newly manufactured devices; or to require greater commitment to refurbishment and repair on the part of device manufacturers, through, for example, requirements on suppliers to offer a certain ratio of refurbished/repaired to newly manufactured goods, with suitable warranties attached to ensure maintenance of standards. In short, producer responsibility legislation needs to be repurposed for the digital age. It cannot remain locked into analogue understandings of firms and manufactured products.

One potential route forward here, and a means to gaining traction with big tech, is to start from the immense wealth that has already been captured (for free) from consumers and sold to third parties. The data consumption that underpins an increasing component of contemporary consumption (think streaming video content of whatever form, web browsing, being on social media) is one of the best examples there is of how digitally mediated consumption creates value out of the surplus – or the trace of our consumption. Google is the supreme example of this.[2] Mostly, the response to this has been to frame it through the political lens, as an infringement of an individual's privacy and in terms of data protection. But what if we also framed this as an economic 'land grab' – as the appropriation of something that people did not know was being turned to value? For that is actually what has happened with the data generated by our digitally mediated consumption. There are any number of examples out there of how this kind of activity, when enacted through more conventional resource extraction, has been retrospectively reimbursed, typically through compensation schemes. Such a fund might be a means to enable the retrospective clean-up of the digital dumps that now litter many parts of the global South.

The emissions burden of experience is an area that takes us directly to the intersection of consumption with the climate emergency. As consumption has become defined as much by doing something as by buying stuff, so it has become increasingly carbon heavy. Less about the embodied carbon in goods, this experience-oriented consumption is reliant on travel, much of it long distance and involving flights or cruise ships. We can see this trend across multiple groups of consumers. No longer confined to the global elite, it extends to the American and European middle classes, to Australians and New Zealanders, to Japanese and to rapidly increasing numbers of Chinese, and to older as well as younger consumers. Tackling this kind of insatiable consumer demand can only be done through climate-related policy, and it is going to require concerted collaborative action at the global scale. Of all the options here, it is consumer carbon budgets that have the most potential. Equivalent to an annual allowance, their roll-out would still permit consumer choice while forcing individual consumers to make trade-offs between different types of carbon consumption (goods v travel/tourism, for example). If connected with established carbon markets, they would allow consumers to purchase further allowances through, for example, participation in other types of activities (for example, volunteering in carbon-accredited schemes such as conservation, restoration or re-afforestation; or contributing either produce or labour to community/urban farming projects). A market in trading credits could also be encouraged, providing financial rewards for low-carbon consumers and incentivizing low-carbon consumption.

Where this takes us is to the need to reconfigure the meta-category of 'the consumer'. Formulated in the period of the Great Depression to stimulate national economies, and unchallenged through the post-war expansion and the high period of globalization, this category is now ripe for revision. My view is that we will not manage to tackle the environmental challenges that are a consequence of consumer demand without decoupling 'the consumer' from the nation state and national economies via GDP. In part, that decoupling was achieved by globalization, as the economies of the global North reaped the full benefit in GDP terms from the purchase of goods manufactured and assembled elsewhere but sold within their borders. But that model is unsustainable. If left to run unchecked, it will end up in the geopolitics of conflict over access to consumer goods and the resources that are critical to their manufacture – a 19th-century resource conflict remodelled for the 21st century through consumption, not production. We see something on these lines starting to take shape already in relation to electric car batteries. So, there is an urgent need to rescale the meta-category of 'the consumer'. We need to do that by aligning the responsibilities of the consumer with responsibilities to the planet. The role of 'the consumer' for the climate emergency is surely to be a global citizen consumer: someone whose pattern of demand is net zero. The challenge is to pull the relevant policy levers to get us to that point.

CHALLENGE 3

If the global waste and recycling industry is part of the global waste problem and not its solution, then how might we disrupt what is now the established fix for waste? And how might discard be economized differently?

The motivations to intervene in relation to waste back in the late 1990s reflect the climate politics of that time, which focused on reducing GHGs. With landfill comprising the third largest contribution to methane emissions, it was inevitable that waste management mitigation efforts focused here, seeking to reduce the flow of material to landfill and particularly to keep biogenic discard out of it. In comparison, incineration (then largely confined to parts of continental Europe, Japan and the US) was acknowledged to be a very minor contributor to CO_2 emissions, through the burning of fossil carbon (plastics and synthetic textiles).[3] This mattered. It meant that the waste management industry's preferred alternative disposal technology to landfill – incineration – could be implemented without further climate scrutiny. The climate politics of the 2020s, however, is defined not by reducing GHGs but by moving to 'net zero'. This has not only moved the goalposts, it has changed the game. For the waste management sector this means: 1) that the pressure remains to reduce landfilling to the absolute minimum; but also 2) that incineration will attract more scrutiny, by virtue of its CO_2 emissions. Although incineration is not a major global CO_2 contributor – unlike say the concrete, steel and shipping industries – it is an emitter.[4] Rapidly growing numbers of large incinerators being built in all regions of the world mean that its contribution is growing (from a small base) while other sectors are already reducing their contributions. Further, as electricity and heating grids are increasingly decarbonized so the spotlight will fall increasingly on energy from waste. A greater share of the fossil-derived energy in the grid will be derived from the fossil carbon burned in energy-from-waste plants. This will become apparent first in those parts of the world that move fastest in decarbonizing their grids. A future with incineration, then, must not only attend to the decarbonization challenge but will also find its future challenged by decarbonization. How might this play out?

There are three broad possibilities here. One is the 'business as usual' option, which would be to regard incineration as necessary to keep material out of landfill. That would require offsetting incineration's CO_2 contribution indefinitely. Currently, there are plenty of other economic actors needing to do the same, but the value of offsetting in achieving 'net zero' remains to be demonstrated. At the very least, a continued reliance on offsetting is likely to bear an increasingly heavy economic cost. The second possibility is to explore the potential for carbon capture.[5] There are major challenges of infrastructural construction and connection here, since most incinerators are sited adjacent to dense urban areas, with limited access to suitable geological formations to act as storage facilities. Third is to address the feedstock itself and specifically, burning fossil carbon. Burning biogenic carbon is regarded as not contributing to global warming, but burning fossil carbon clearly does.[6] Yet, fossil carbon is critical for energy conversion efficiency, which means that within the industry, carbon capture is seen as the only realistic technical option. Step beyond the waste management industry, however, and increasingly very different futures for fossil carbon are being imagined. These matter because they show that a fossil carbon feedstock for incineration in its current form cannot continue to be assumed.

So there are major challenges ahead. But overarching these is the challenge that has always faced the waste management industry, namely that it is a lagging industry. Its inputs are the effect of, and determined by, the innovations, designs and manufacturing investment decisions of previous decades. As a result, innovation in the sector has been conducted in isolation from other areas of the economy, and an effect of that is that it has been largely technical and 'kit' focused, driven primarily by efficiency gains. My contention is that the climate emergency and push to decarbonize economies requires that this change. Addressing decarbonization means the waste management industry can no longer sit at the end of the chain. Rather, the industry needs to be reconfigured such that it integrates with the requirements of increasingly decarbonizing production, while maintaining the capacity to manage wastes of negative value. This is likely to require the development of a focus on managing discard, not managing discard to become waste.

Making space for a discard management sector

Managing discard is about maintaining openness for revaluation. It entails the collection of materials that are recognized to retain an inherent value for what they currently are. It is not about ensuring that what is discarded becomes waste. As such, managing discard is a fundamentally different business to the model that currently orders the waste industry. It is a business that will need to be ordered on the collect-curate model. When we look at some of the more agenda-setting thinking in chemistry – the academic discipline that sits at the heart of decarbonization in a material sense – we can see that careful collection and storage (curation) is already being recognized as a requirement for a chemicals industry for a decarbonized age. What is being emphasized here is not collecting oil-based feedstock to feed back into further oil-based production (the 'downcycling/recycling' approach discussed in Chapter 5). Neither is it the 'transform wastes into innovative products' model promoted by circular economy thinking (Chapter 5). Rather, foresight thinking on decarbonized and decarbonizing chemistry is foregrounding the requirement to collect different types of feedstock to those currently collected (for example, agricultural wastes); the need for chemistry to move away from the development of proprietary polymers and to focus on an agreed core of baseline non–carbon-generating feedstocks (which will need to be recovered); and the requirement to collect and curate (not burn) the plastics that will emerge from various industrial uses in the next 25 years.[7] Doing that will require either: 1) a fundamental reconfiguration of the existing waste management sector, to move away from established technologies of burning, to embrace the collection and curation of multiple new materials; and/or 2) the development of a new discard management sector.

The current positioning of the waste management sector as the economic end of the line is a significant barrier to the development of discard management possibilities within the sector. It is likely that connections are going to need to be made, or at the very least enabled, by policymakers. This is because not only are the chemicals and waste management industries ones that do not easily come together, but also connections made within and between industry will likely reproduce existing lagging effects, as the vested interests of petrochemicals shape a response from waste management that allows for the perpetuation of the carbon-based model. Instead, waste management needs to be incorporated into the many academic and early-stage innovation conversations occurring in relation to the decarbonization agenda, and now, from the outset, not left as an afterthought or brought in as a 'tack-on'. In particular, waste management needs to hear foresight thinking in chemistry. But what else might be done to stimulate the development of discard management more broadly? I have three suggestions.

The first, and perhaps most controversial given the vested interests of waste management, is to dispense with the waste hierarchy. There is no doubt that the waste hierarchy has been a useful heuristic in bringing what happens with and to waste back into people's minds, but as this book has shown, it has had unfortunate unintended consequences. It works to underscore and support sunk capital investment, as the capital plant now on the ground is a materialization of decisions that were guided by the hierarchy. The question that should now sit uppermost in considering the future value of the waste hierarchy as a policy device is how well it drives moving in the direction of decarbonizing economies. And the answer there is that it doesn't. The only elements of the hierarchy that align closely to this agenda are reuse and waste prevention strategies, neither of which have been prioritized given the focus the hierarchy permits on recycling and energy recovery as better options than landfill. Dispensing with the hierarchy, while continuing to recognize that landfill remains the destination of last resort, will give the space to pay greater attention to formulating policies that actively encourage reuse, while not rewarding recycling as a universally 'good thing'.

Secondly, there is a need for data, and specifically data on: 1) the material stocks embodied in economies, and projected stocks, on varying timelines, covering not just the goods held in households but those held in buildings, infrastructure and manufactured goods; and 2) stocks of new feedstocks likely needed to supply a future chemical industry. Data on agricultural wastes is particularly important here and may signal an urgent need to reorient the direction of travel, which has focused on identifying technical solutions to particular agricultural wastes per se rather than the whole system change now being envisaged in reimagining chemistry for the mid-21st century.

Thirdly, an emphasis on curation will require new land and infrastructure for materials storage. Likewise, collecting new materials will also require new infrastructure. It is unlikely that such a degree of fundamental restructuring can be accomplished by relying solely on the market. Both state financial support and integrated economic planning will likely be required to bring about this degree of change.

Rescaling waste management solutions to encompass discard management

A further challenge that follows from the foregoing is that careful collection and curation requires a different scalar solution to that which underpins capital-intensive waste management. Rather than producing large volumes of roughly sorted materials, careful collection requires that material distinctiveness is maintained in and through collection and transportation, while curation requires that materials are stored in the manner of an archive – that is, catalogued, ordered spatially in ways that recognize material distinctiveness, and with efficient systems of search and retrieval to support user demand. It would be unrealistic to expect the waste management sector in the form it currently takes in the global North to have these types of expertise. But, there are models of good practice.

The best models for the careful collection of discard are the Indian cases discussed in Chapter 6, and then, in relation to curation, the heritage sector has long been the leading exponent of archival and curatorial best practice. The Swachh Bharat interventions outlined in Chapter 6 are very much solutions that have been designed in and for the global South; they recognize the importance of waste as a livelihood and they are heavily labour intensive. This kind of model will not transfer lock, stock and barrel to the economies of the global North. So how else might collection become careful? I think that here we have to try some 'left-field' thinking and think 'what else comes to households or businesses with intensity?'

The answer to that question is the burgeoning demand for 'last-mile' logistics that has been an effect of the switch to online commerce. Much as the globalization of waste rested on seeing the capacity in the back run of global shipping, so too – I would suggest – there is a back run that could be exploited to capture discard. This is 'last-mile' logistics. All those deliveries are currently one-way only. With small adjustments to last-mile delivery vehicles, they could also become micro collection vehicles, at no extra transport cost. With growing numbers of people homeworking, the challenges of coordinating handover/exchange (and the burden that places on delivery workers) would also be ameliorated. For those who do not work from home, local, neighbourhood-based collection points could serve the same purpose. Local logistics hubs – the deconsolidation points for last-mile delivery – could be expanded to serve as small-scale materials collection hubs, or consolidation points, where the first stages of careful discard amalgamation might proceed. At the same time and again taking lessons from Swachh Bharat, those same venues might act as or supply outlets for used goods.

Such arrangements would likely only work in relation to 'dry' discard, or the non-organic fraction of household discard, for example, stuff currently excluded from household waste collections in the global North, like the plastics that the waste management industry will not collect, or small items such as batteries and shoes and textiles. 'Wet' organic discard, such as the leftovers from food preparation and meals, would not be suitable for such collection arrangements, but again there is the possibility of bolting this kind of collection onto another form of last-mile logistics – the burgeoning market in meal delivery. Local delivery 'riders' might also become local collection 'riders' for organic leftovers. If urban city farming is to take off – as it must in a decarbonized future – then micro-scale, intensive urban cultivation is vital. That will require compost available at the neighbourhood scale and likely, too, small-scale biogas production facilities, of the type that can heat community-scale polytunnels. In other words, household food waste needs to be rescaled to the neighbourhood-community level, to become the stuff that enables local food production. Declaring it to be already waste and collecting that at scale, to produce biogas, may be the waste management industry's preferred route forwards with food waste, but such perspectives need to be challenged and evaluated in relation to their fit for purpose in shaping resilient futures for all communities.

★ ★ ★

Running across all three sets of challenges are what I would identify as the three Ds that underpin reframing the global waste problem. These are decarbonization, digitally mediated worlds and discard. For too long the global waste problem, and thinking about waste, has been stuck in a silo that is the product of linear thinking in an exclusively offline world. The current fix to the global waste problem perpetuates that; as I hope this book has shown, it's a capitalist fix to the problem. What we have created, in endeavouring to solve the problem of 'too much waste', is a situation that pushes discard to become waste and that then turns waste to value – for capital, in the form of waste management transnationals, and for finance capital.

The challenge ahead is to reposition the global waste problem within the climate emergency, and to see the current fix to global waste as part of that wider challenge. My belief is that the solutions to the global waste problem must be thoroughly integrated in decarbonizing and decarbonized economies; they lie in seizing the opportunities, as well as the challenges, of increasingly digitally mediated futures; and, ultimately, they are to be found by capturing discard back from the category of waste. Realigning and rewiring the waste problem through the climate emergency offers the opportunity to bring climate goals and waste goals back in the same space, and to make the space for a more careful, curatorial approach to resource utilization. It will not, and never can, eliminate the category 'waste', but what it will do is to make that category much more narrowly defined – an important but much more precisely engineered solution for the management of particular wastes, rather than a means to harvest the detritus of a world driven by ever expanding and ultimately unsustainable consumption.

Notes

Preface
1. Gille and Lepawsky (eds 2022).
2. Karen Hansen's work on Zambia was the trail blazer here. See her *Salaula* (2000).
3. O'Neill (2019).
4. My use of this term goes beyond that of the everyday noun – of a process that, once begun, sets in train a whole host of complex problems, which often spell the world of trouble. It also cites some of the central elements of the Pandora myth. In Greek mythology, the opening of Pandora's box, or casket, lets loose on humanity an array of harms and evils, while leaving hope trapped in the box. The myth is one of the most appropriate metaphors for thinking about the activities of resource recovery and recycling. Discarded things can appear to offer the gift (or even hope) of boundless, endlessly recoverable resources, but they require the work of transformation to do that. That transformation might involve manual or mechanical disassembly; alternatively, it might involve a chemical or biological process. Regardless, it often lets loose materials that are toxic to life forms.
5. Alexander and Reno (eds 2012).
6. See, for example: Hird (2012); Hecht (2018).
7. See, for example: Lepawsky and McNabb (2010); Lepawsky and Mather (2011).
8. See, for example: Hird (2021); Gille and Lepawsky (eds 2022).
9. Projections for demand for lithium for electric vehicle batteries, for example, show an 11-fold increase in the 10 years between 2020 and 2030. See: Visual Storytelling Team in London (2021). For an overview of an increasingly lithium dependent world, see: Sanderson (2022).
10. On recovering and recycling EV batteries, see: Harper et al (2019).
11. Government Office for Science (2017). GO-Science produces two annual two-part reports that give scientific evidence and policy recommendations on a chosen area of UK policy.
12. Abramovitz et al (2015); Abramovitz (2017). See too: Ebola Response Anthropology Platform – ebola-anthropology.net.
13. See the various contributions to Gille and Lepawsky (eds 2022).

Chapter 1
1. That said, there are limitations to waste data that need to be recognized. The world over, waste data, whichever category it is reported in, is – as my colleague Mike Crang loves to express it – 'deeply dodgy data'. Aside from the non-existence of data for some countries, the challenges include inconsistencies in definition and the absence of standardized categories across countries, making comparisons difficult; differences in measurement such as whether waste is weighed 'wet' or 'dry' and at the point of generation or the point of collection; the widespread lack of weighbridges in lower-income countries, which means that large amounts of waste go unmeasured and unaccounted; and uncertainties

NOTES

over estimation, as well as over estimation techniques. On top of this, the incentives to under-reporting and misreporting are considerable. This is because waste, outside the waste management industry, is a cost. The most obvious instance of this occurs in relation to the available data on the trade in wastes, where illegal exchanges, purposeful miscategorization and under-reporting are less the exception and rather more the norm in business practice – see Gregson and Crang (2015); Neuwirth (2011). At best, then, when we look at any type of waste data, a key maxim is to exercise caution. The best approach is to regard data as estimates, in some cases heroic, and be mindful that inferences may well be misleading.

2   In New Zealand, an estimated 50 per cent of total wastes come from construction and demolition (Inglis, 2007 cited in Akhtar and Sarmah, 2018).

3   Akhtar and Sarmah (2018).

4   There is a considerable literature that addresses the legacy of the nuclear industry in radioactive waste. Blowers (2016) provides an excellent overview. For a powerful account of the challenges of managing these wastes, written from inside the UK industry, see: Bolter (1996). A more expansive reading of the wastes of the nuclear industry comes in the work of Gabrielle Hecht, whose research examines the entanglement of the residues of uranium mining in Africa with miners' bodies and the fabric of the built environment. See: Hecht (2018). The challenge posed by radioactive wastes, at least in democracies, is to find a means to their management that is not only scientifically and technically safe, but also socially and publicly acceptable, across generations and geological time. Across governments and within the nuclear industry, that task is widely acknowledged to have required a shift from top-down, 'decide, announce, defend' governance approaches to a deliberative and participatory approach. A sizeable academic literature addresses the deliberative/participatory turn in nuclear waste governance, typically from a within-country or comparative perspective. For examples see: Chilvers and Burgess (2008); Bickerstaff (2012); Bergmans et al (2015); di Nucci et al (2017). Notwithstanding the effort, currently there are only two countries in the world to have reached the point where a deep geological disposal facility for radioactive wastes is at an advanced stage: Finland and Sweden. Everywhere else, these wastes remain in interim storage, sequestered within facilities that were only ever designed to be, at best, temporary solutions to the radioactive waste problem.

5   For examples of the kind of ordinary materials categorized as low-, or very low-, level radioactive wastes, see: NDA (2019, Table 2 p 21).

6   'Higher activity' radioactive waste is the term that is applied to high-level and intermediate-level radioactive wastes in the main, distinguishing these from low and very low-level radioactive wastes. Baseline inventories provide estimates, by country, of the volume and radioactivity of the wastes produced by the nuclear industry. An example is the UK's Radioactive Waste Inventory. From that inventory, 2019 data lists 133,000 m^3 of radioactive wastes, but the lifetime projected arisings from the decommissioning of the UK's existing nuclear estate (that is, minus any 'new build') are 4,560,000 m^3. Currently, low-level and very low-level radioactive wastes account for 21 per cent of the UK's inventory by volume (28,440 m^3). Projected arisings have the equivalent figures as 4,310,000 m^3 or 96 per cent of the inventory (www.ukinventory.nda.gov.uk). Low-level wastes also comprise over 90 per cent by volume of the radioactive wastes produced in Europe (Garcier, 2014).

7   They demand that the nuclear industry find ways to authorize the wider circulation of materials whose radioactivity hitherto has been the means to restricting circulation to a few tightly prescribed within-industry movements. See: Garcier (2014).

8   Notwithstanding the immense challenges of measuring it, the World Bank estimates that industrial wastes (or wastes emanating from industrial activity) exceed consumer wastes by a factor of 18. See: Kaza et al (2018).

9 Clapp and Swanson (2009, p 317).
10 Single-use plastic shopping bags became a habitual part of grocery shopping in much of the world from the 1980s. Unlike most instances of global environmental policy intervention, action relating to plastic bags began in the global South, in Bangladesh in the early 1990s. It then spread to other parts of South Asia (India, Nepal), to Taiwan and South Africa. However, it was not until the mid-2000s that efforts to reduce plastic bag consumption started to become widespread in the global North, through a mixture of voluntary codes for grocery retailers, taxes on bag use such as the 5p charge in the UK, and bans on use, such as in Ireland. See Wagner (2014) for a US example. Clapp and Swanson (2009) associate this pattern of South-North interventions with the strength of the plastics industry in the economies of the global North.
11 Hoornwey and Bhada-Tata P (2012).
12 Kaza et al (2018).
13 'Enough' in this instance relates to its 'sufficient' meaning (Coyle, 2011). Interest in this covers a broad church. It includes the likes of the New Economic Foundation (neweconomics.org), arguing for the substitution of the growth model with a well-being model, allied to shorter working weeks, a guaranteed income for all and the expansion of local and community-based economies. Then there is work in the economics of happiness canon, where the argument is made that peak happiness appears to coincide with relatively modest levels of income. Known as the Easterlin Paradox – after a 1974 article by the economist Richard Easterlin – that 'sweet spot' was quantified at an average of $15,000 per person per annum by Richard Layard in his 2005 text, *Happiness*.
14 See, for example: tristramstuart.co.uk and feedbackglobal.org
15 Gregson (2007).
16 Perhaps the most prominent example of this is Olio (olioex.com). Initially conceived as a way by which the food retail and hospitality sectors could be connected with food charities, these apps are now seen as a means of connecting all with spare food to anyone else who might want it. They therefore encompass people living in streets and neighbourhoods, or ordinary consumers like you or me. Their promise is to save food from becoming waste and to turn us all into waste-saving citizens.
17 Ending global food poverty is the second of the UN's sustainable development goals. In 2020, an estimated 2.37 billion people (roughly 25 per cent of the world's population) were without food or unable to eat a balanced diet on a regular basis (sdgs.un.org). Food poverty research in the global North is primarily focused on food banks and food charity and their connection to austerity; on food bank usage by households dependent upon benefits and/or precariat work; and on the dietary deficiencies that characterize food bank dependency. See: Tarasuk et al (2014); Booth and Whelan (2014); van der Horst et al (2014); Bazerghi et al (2016); Cloke et al (2017).
18 In the UK, WRAP (the Waste Resource Action Programme) estimate that the average UK household wastes 165 kg of edible food per annum (or the equivalent of £500), with each person estimated to be wasting 69 kg (or their own body weight) – or £210.
19 Households constitute the biggest source of food waste in the UK, accounting for 70 per cent of the post-farmgate total (6.6 million tonnes in 2018, compared with 8.1 million tonnes in 2007, when data was first available). See: WRAP (2020a).
20 Castrica et al (2018); see too: Bloom (2011).
21 In 2020, in response to China's growing food waste problems, Xi Jingping declared 'waste is shameful, thriftiness is honourable'. 'Operation Empty Plate' was declared. This seeks to reduce hospitality food waste. It limits a group of n diners to n-1 orders and has banned 'mukbang' videos (a Korean innovation), which celebrate binge-eating. Food operators face fines of 10,000 yuan (>£1,000) for failing to comply with the new legislation and their clients face charges for 'excessive' leftovers. See: Davidson (2020).

NOTES

22. In the UK, the best example of this is Fareshare (fareshare.org.uk). The organization's origins lie in a partnership between the homeless charity Crisis and the UK supermarket Sainsbury's. In 2004, it became an independent charity and is now the national network for redistributing surplus food from all parts of the food industry. The formal partnership is with 18 independent organizations, including three of the major supermarkets (Asda, Sainsbury's and Tesco), the Co-op, Kelloggs, Nestlé, Birds Eye, Coca Cola, Greggs, Tropicana, Fowler Welch and General Mills. Over 500 companies from across the food and drink industry have participated in the network, and it serves more than 10,000 charities and community groups, including school breakfast clubs, older people's lunch clubs, homeless shelters, food banks and community cafés.

23. That sheer unpredictability is nicely illustrated by research that talks about the challenges of making meals from stuff like piles of avocados and chocolate mousse. See: Alexander and Smaje (2008).

24. There is a small but significant sociological literature on food waste, mostly based on research with UK households, that I draw on here. It highlights how food waste emerges from the rhythm and pattern of contemporary everyday lives and their intersection with domestic technologies such as the fridge and freezer, as well as concerns over food safety. See: Evans (2012); Watson and Meah (2012); Evans et al (2012); Evans (2014); Hebrok and Heidenstrom (2019). An indication of the importance of sociological context to understanding food waste came with the COVID-19 pandemic: with more available time under 'lockdown', more UK households engaged in cooking preparation from scratch, while anticipated food scarcity and an unwillingness (or highly restricted capacity) to engage in food shopping led to more careful food provisioning, storage and use within households. Tellingly, with the relaxation of restrictions and increased mobility, the amounts of food waste being generated increased again. See: WRAP (2020a).

25. There is a welter of sociological literature that has examined patterns of women's employment since the mid-20th century and its effects in, and on, domestic work. From early on, a key concern was to document the effects of domestic technologies on the gender division of labour in the home, on the amount of domestic work being done, and on the relationship between paid and domestic work as this was felt in women's lives. Key texts include: Schwartz-Cohen (1985); Gershuny and Robinson (1988); Cockburn and Ormrod (1993); Cockburn and Furst-Dilic (eds 1994); Hochschild (1989); Bittman et al (2004). Time emerged as a major theme from this research, but with differences in how it was understood. Whereas for some women, time was understood as time used or spent, for others it was time as felt, or lived, which for women in employment led to feelings of time pressure, and the sense of what is termed 'the double burden' for women.

26. The phrase is Ray Pahl's, from his 1984 *Divisions of Labour*, which contrasts 'time-poor' households – in work and often with two working adults – to 'time-rich' ones – mostly unemployed, for whom 'getting by' is a mode of everyday life.

27. Calvignac and Cochoy (2016).

28. One of the best examples of this is a pattern of meals that was commonplace in many working class households in England post 1945: Sunday, hot roast lunch; Monday, cold meat dinner; Tuesday, soup broth dinner made from the meat carcass, bulked out with root vegetables; and with the rest of the meat being used for sandwich fillers. By the end of the week nothing was left of this meat.

29. WRAP (2020, 2020b Appendix 1 p 15) provides an indicative list. Every day in the UK sees 20 million slices of bread thrown away; 4.4 million whole potatoes, 0.9 million whole bananas, 1.2 million whole tomatoes, 2.7 million whole carrots, 1 million onions, 0.8 million whole apples and 0.7 million whole oranges, while 3.1 m glasses of milk are poured down the sink.

30. While the veterinary literature highlights that there is little difference in meat quality between the carcasses of pigs fed on a grain diet and a food waste diet, biosecurity in food waste processing can be a challenge. The most recent high-profile example of biosecurity in relation to the feeding of food waste to pigs is the 2001 foot and mouth outbreak in the UK (see: Alexandersen et al 2003). This led to a ban on these practices in the UK and the EU in 2002. In contrast to the situation in the EU and the US, in South Korea and Japan, over 40 per cent of food waste is used for animal feed (Dou et al 2018). A rather different challenge to the biosecurity one comes from the agrifeed business, where the inherent variability in the calorific content of food waste as an input, and the high costs of separation, puts such a product at a disadvantage in a market dominated by standardized grain-based products designed to optimize meat production on the bone.
31. Naming is the key distinction between animals kept as livestock and companion animals. Whereas the latter have names, the former are identified by numbers.
32. Typically used by older people – old enough to remember these ways of feeding pets – it's a generally derogatory term, vernacular for a mess.
33. Zimmerman (1951).
34. Roosevelt et al (1991). See also: Needham and Spence (1997). On deposition, discard and their connection to pits, see: Garrow (2012); Thomas (2012); Jervis (2014).
35. Miller (ed 1995).
36. Miller et al (1998).
37. Miller (1998).
38. Daunton and Hilton (eds 2001); Cohen (2003); Kroen (2004); Trentmann (2006, 2016).
39. Warde (1999); Southerton and Shove (2000); Warde (2005); Southerton (2006).
40. Thompson (1979).
41. There is a significant body of academic research on Wedgwood, as befits the lead firm in the English potteries industry in the 18th century. While Wedgwood introduced novel forms of factory-based production, their most important innovations were on the demand side, particularly in relation to sales and distribution. See: McKendrick (1960); McKendrick et al (1982); Dodgson (2011); Holt and Popp (2016).
42. On collecting generally, see: Belk (1995). On stamp collecting, see: Gelber (1992); Brennan (2018).
43. One of the classic texts in cultural studies uses the Sony Walkman as its case. In their revised edition, du Gay P et al (2013) frame this iconic artefact through the notion of 'the circuit of culture' and also through the wider relationship of production to consumption, and the reworking of that relationship through digital technologies (known as produsage). This overlooks that Sony Walkmans, as devices, have mostly become rubbish value, and the consequences of that trajectory in waste. A different strand of work in cultural studies, inflected through what is known as 'the material turn', has tackled this blind spot head on. Instead of focusing on iconic objects, it focuses on the mundane objects and materials that now charge environmental activism and politics, notably plastic bags, bottles and packaging. See: Gabrys et al (eds 2013); Hawkins et al (2015).
44. Hawkins et al (2015). It is no coincidence that the use of these goods affords consumers the capacities of convenience, hygiene and freshness, often in conjunction. Plastic drinks bottles do this; so too does plastic packaging in relation to fresh produce, while the value of disposable personal hygiene goods lies in their capacity to be both absorbent and disposable, thereby removing the need for humans to do the work of dealing with and cleaning dirty, messy material (make-up, blood, excreta) from other materials.
45. Gregson et al (2007b).
46. On China, see: Edmonds (ed 2000); Economy (2004); Lora-Wainwright (2013); Lora-Wainwright (2021). On Russia and the former Soviet Union, see: Henry and Douhovnikoff (2008); Kochetkova and Pokid'Ko (2019).

NOTES

47. Hardin (1968).
48. There is a substantial literature on waste scavengers in developing countries. See: Medina (2007) for a useful entry point.
49. See the literature on dumpster diving and freeganism: Ferrell (2006); Edwards and Mercer (2007); Barnard (2011, 2016); Edwards and Mercer (2012). See too: Lane (2011).
50. See: Rathje (1984); Rathje and Murphy (1992).
51. Storage is one of those distinctly niche aspects of economies that barely figures in analyses but that is integral to their functioning. This is because storage provides: 1) buffering: it is the means to manage flows (of goods, services, data) and thus to balance supply and demand. Examples of this kind of storage include the tax-exempt commodity warehouses that are to be found around global ports, manufacturers' and wholesalers' warehouses and their retail equivalent, the distribution centre. Storage is also: 2) the means to manage overflow in economies.
52. Methane is the second most abundant GHG after CO_2, accounting for ~20% of total emissions. It is estimated to have a 100-year global warming potential 28–34 times that of CO_2, while its 20-year potency is 84–86 times that of CO_2. Landfill is the third largest contributor to global methane emissions (estimated to account for 16 per cent of the total) after fossil fuels (33 per cent) and livestock production (27 per cent) – epa.gov; unece.org.

Chapter 2

1. Discard studies is concerned less with waste as the primary object of study and more with 'the wider social, economic, political, cultural and material systems that make waste and wasting the ways they are' (www.discardstudies.com). As an overtly 'critical' field of study, discard studies interrogates and questions what is taken for granted as 'normal' and normative, focusing on the power relations that constitute, and benefit from, these arrangements.
2. The significance of discarded food for human history is considerable, since this was a key means to species domestication. It led to human cohabitation with today's primary companion animals: dogs and cats. As ecological scavengers, wolves and then dogs were attracted to discarded food, as too were rodents; cats, as predators, were admitted into human settlements as the means to control rodent populations. For an accessible introduction, see: Ault (2015).
3. Analyses of the stomach contents of birds and animals seek to establish the scale of plastic ingestion by other species. As the analysis of microplastics has accelerated, it is establishing the extent of microplastic presence, in humans as much as other life forms. For a review with respect to sea birds, see: Provencher et al (2017).
4. See: Boulding (1966); Georgescu-Roegen (1971); Frosch and Gallopoulos (1989).
5. Ellen MacArthur Foundation (2012, 2013, 2014).
6. Washburn (1960).
7. There are varying definitions of the Anthropocene and differing views on its origins. For useful entry points to the scientific literature, see: Steffen et al (2007); Steffen et al (2011); Lewis and Maslin (2015); Waters et al (2016). By contrast, Gabrielle Hecht sees the Anthropocene as the apotheosis of waste, and as resting on a huge expansion in the quantity, extent and durability of discards (Hecht, 2018). In contrast to scientific accounts, she argues that the effects of this are not just planetary but that the Anthropocene has different manifestations depending on where one is in the world. For Hecht, then, there is no single geological epoch but rather multiple Anthropocenes to be uncovered.
8. Marzke (1997, 2013); Tocheri et al (2008). For a useful review, see: Kivell (2015).
9. Douglas (1966).
10. See: Liboiron (2019, 2022).

11 Douglas uses a couple of mundane examples to illustrate what she means by dirt, neither of which is literal. They are shoes on a dining room table and used cooking utensils in a bedroom. What matters about these things is not the things but rather their situation. So, there is nothing about shoes or cooking utensils per se that renders them dirt. Rather, it's their placing that is the contravention of notions of order and that makes them 'dirty' or defiling. In this way, Douglas' arguments about dirt link to her wider concerns with pollution and taboo.

12 Liboiron (2019, 2022).

13 That same quality of coming to notice underpins the vital materialist strand of thought in the social sciences, where it has been used as the prompt to develop a stronger attentiveness to what is termed 'thing power' or material agency. See: Bennett (2004).

14 Beall (2006 p 83).

15 Butt's (2019, 2020) work on Lahore is exemplary in this regard. He shows how colonial administrators, seeking to make sense of caste, cemented associations between certain people and the handling of particular materials, notably the Chuhra (sweepers and scavengers). Gill's work on recycling work in India also highlights the continued effects of caste, through the maintenance of distinctions between pickers, dealers and buyers in the recycling trade (Gill, 2007).

16 See: Fahmi (2005); Kuppinger (2014).

17 See: Harper et al (2009); Skobla and Filčák (2020); Dunajeva and Kostka (2021). The long association of Roma with waste picking (see: Rosa and Cirelli, 2018) has been the basis for their wider stigmatization and exclusion. However, it is the inability to pay for privatized waste collection services, subsequent indebtedness and then their exclusion from waste collection services that has resulted in piles of rubbish accumulation in Roma settlements across Central and Eastern Europe. The effect has been to reframe the Roma as 'dirty Roma'.

18 Gregson et al (2016).

19 For a comparative summary, see: Warde et al (2007).

20 This general tendency has given rise to a new paradigm in sociology called 'mobilities'. The core argument here is that sociology needs to be reworked to reflect the heightened mobility, speed and intensity of modern life. See: Hannam et al (2006); Sheller and Urry (2007); Urry (2007), as well as the journal *Mobilities*.

21 The drive-thru emerged in the US and is closely associated with auto-culture and auto-mobilities. On auto-culture and automobilities, see: Miller (ed 2002); Dant (2004); Thrift (2004); Urry (2007). The first drive-thru outlets appeared in the US in the 1920s, but it was in the 1950s in California that they became increasingly part of US driving culture. Artefacts from this period (for example, menus from the In-N-Out burger chain) are in the National Museum of American History, which also includes a short online synopsis of the drive-thru in the US – www.americanhistory.si.edu. In 1975, the first McDonald's and Burger King stores opened. By the late 1980s in the US, McDonald's was making most of its sales from drive-thru outlets. The phenomenon is now worldwide. An indication of the degree to which fast food drive-thru outlets can become a major part of the culinary landscape is provided by the UK. A mapping exercise of drive-thru outlets identified 461 McDonald's and 228 KFCs, with densities at their highest around many cities in Northern England and the western Central Belt in Scotland. Glasgow and Manchester each have 48 outlets; there are then 41 in Birmingham, 37 in Liverpool, 34 in Sheffield, 32 each in Leeds and in Newcastle (Drive-thru density in the UK – www.zuto.com).

22 See: Schlosser (2002).

23 Even back in the early 2000s, when waste policy was in its infancy in the developed world, there was mounting concern about the discard associated with fast food. A Defra report from 2004, based on research by Keep Britain Tidy, highlighted the growth in litter

derived from fast food and drink (Defra, 2004). More recently, the rise of home-delivered fast food is leading to increased amounts of this discard in household discard. A study of packaging waste from food delivery in China estimates that the first half of 2017 saw 4.6 billion orders, of a value of $28 billion. It also estimates that this grew from 0.2 million tonnes in 2015 to 1.5 million tonnes in 2017, generating 1 per cent of China's municipal solid waste (Song et al 2018).

24 Disposable cups are manufactured to allow consumers to carry hot drinks safely. Safety is conferred in two ways: through a sealable lid, with an opening to sip the drink, and through the composition of the cup itself. The cup is made of paper combined with a polyethylene inner layer, with the two surfaces fused through a heat treatment. This makes for a strong, watertight cup but also a cup that the waste management industry labels as 'difficult' to recycle (Solving the disposable cup conundrum – www.livingcircular.veolia.com). 'Difficult', however, often translates to mean technically possible but economically challenging.

25 One example of this is the roll-out of a collection system for disposable cups by the waste management firm Veolia. This was developed first through a partnership with Costa and is now being expanded to include businesses and train stations (Solving the disposable cup conundrum – www.livingcircular.veolia.com). Notably, this collection system is still located inside premises; it is therefore targeted more to on-premises (that is, sit-in) 'on-the-go' consumption. Rather different is the environmental charity Hubbub's smart/digital innovation, currently rolled out to Leeds, Swansea and Edinburgh. This relies on an app that consumers need to access to find the nearest bin to recycle their disposable cup.

26 Complaints about drive-thrus and littering are widespread. See, for example: Litter explosion after opening of drive-through McDonalds in Inverness www.pressandjournal.co.uk.

27 One of the clearest examples of the connection in the UK came post the first wave of the COVID-19 pandemic. Having been in 'lockdown' for the spring of 2020, controls began to be relaxed in May, including for drive-thrus. The result, following a period of no such litter, was a huge increase across the length and breadth of the country (see: www.kentonline.co.uk; www.dailypost.co.uk; www.walesonline.co.uk; www.cambridgeindependent.co.uk. This led to a renewal of calls for the identification of litterers through recording car registration plates on drive-thru purchases.

28 Kristeva (1982).
29 Veblen (1994).
30 Malinowski (1922).
31 Dietler and Hayden (eds 2001).
32 GWP Group (2020). An indication of the extent of additional household expenditure on food and drink at Christmas in the UK is given by Pitts et al (2007).
33 See, for example: Diwali food menu for lunch and dinner party (timesofindia.indiatimes.com); Diwali 2020 food: delicious Diwali food menu for a memorable lunch and dinner party (food.ndtv.com); Gowardhan M (2020); Eid 2020: full Eid party menu with recipes to make your Eid-al-Fitr celebrations delicious (food.ndtv.com); Afreen's Kitchen nd.
34 This would include multiple types of fish, dumplings, spring rolls, glutinous rice cakes, sweet rice balls, longevity noodles and fruit – Chinese New Year food: top 7 lucky foods and symbolism (www.chinahighlights.com).
35 There is a considerable feminist literature, much of it inspired by Foucault, on body thinness and dieting. See: Bordo (1993); Heyes (2006).
36 See: Christmas food waste: over 4 million festive dinners are thrown away each year – www.independent.co.uk. Quoting figures from Unilever, this article itemizes the full wastage as an estimated 263,000 turkeys, 7.5 million mince pies, 740,000 slices of Christmas pudding, 17.2 million Brussels sprouts, 11.9 million carrots and 11.3 million roast potatoes.

37 Caplow (1982).
38 See: The Sun: The Morning After – www.thesun.co.uk; BBC – Glastonbury 2019: Festival clean-up underway – www.bbc.co.uk; www.glastonburyfestivals.co.uk – Our waste policy; Jackman J (2020) Glastonbury Festival's Carbon Footprint – www.theecoexperts.co.uk.
39 Coveya (2019); Cleaning up the waste at Glastonbury – www.expertskiphire.co.uk. In this way, 1,000 tonnes of discarded materials – 50 per cent of the 2019 total – was captured for recovery. It is important to acknowledge that such rates of recovery rest on extremely high levels of voluntary labour. The amount of work this entails, and particularly the time it takes to separate out recoverable material from waste, is also instructive as to why treating 'on-the-go' discard in ways that maximize materials recovery is so challenging. Unlike Glastonbury, organizations like train companies and local authority parks have neither the time nor the financial headroom to put in place materials recovery infrastructure and to engage anything like this amount of labour.
40 It is worth noting that this figure is minus any charge for the immense amount of 'picking' labour that goes into the clean-up. Voluntary labour is precisely that, voluntary. If one were to cost this labour, even at minimum hourly rates, the costs would double.
41 See: Gregson et al (2007a); Gregson (2007).
42 Gregson (2007).
43 GWP Group (2020). An estimated 25 per cent more waste is produced by US households in the period from Thanksgiving to New Year.

Chapter 3

1 See: Schumpeter (1942). Although the term creative destruction is identified closely with Schumpeter, its roots in German thought stretch in an arc from Nietzcshe (who in turn drew on Indian philosophy) to the economist, Sombart – see: Reinert and Reinert (2006). More recently, the centrality of creative destruction to mainstream understandings of economic growth has been highlighted by Berman (1987) and by Ricci (2020), whose manifesto for political science urges that, to counter the dominant narrative of economics, it is imperative for the other social sciences to highlight the destruction that is integral to creative destruction, and its effects. Ricci states: 'without mincing words, the phrase "creative destruction" gently spins an occasional brutal process that in plain English is analogous in some respects to war'.
2 See: Dicken (2014).
3 There is a vast academic literature focused on these processes, most particularly with respect to the recession of the 1970s and early 1980s; on their effects on local, regional and city economies, and socially; and on subsequent interventions that sought to promote alternative employment futures, both at the policy level and community-led initiatives. At the general level, a useful starting point for the US is Bluestone and Harrison (1982), while Massey (1984) offers a UK comparison. A welter of case study-based research then gravitated to places that were particularly hard hit by the capital shifts going on in the extractive and manufacturing industries up to and including the early 1980s. See, for example, Cooke (1989). In the US, Detroit has long stood as the 'poster child' of this process, while in the UK, regions such as the north-east of England (particularly Teesside and the former coalfield communities) attracted considerable attention.
4 See: Ford (2015, 2021); McKinsey (2017); Acemoglu and Restrepo (2020).
5 Piketty (2014).
6 Household bin studies focus on characterizing the material composition of what is thrown away by consumers and, along with the excavation of landfills, they are at the heart of what is termed 'garbage archaeology'. The term was instigated by William Rathje's Garbage Project, which analyzed nearly 200 tonnes of fresh garbage taken from over 20,000 households across seven metropolitan cities in the US, along with 45.3 tonnes (dated from

NOTES

1920 to 1991) from 19 landfills and open dumps associated with 15 North American cities – see: Rathje and Murphy (1992). This work has highlighted transformations in the material base of society through the 20th century, specifically: the decline in coal as a means to domestic heating; the rise in plastic and growing volumes of food waste – see: Rathje (1984). It is mirrored in its findings by parallel work in New York, see: Walsh (2002).

[7] In practice, the two are connected through a range of arguments that link the visibility of waste in the wider environment (or the 'too much waste' problem, as evidenced by waste statistics – see Chapter 1) to overconsumption. They then infer from that the profligacy and wastefulness of modern consumers, as compared with their historical counterparts, who are typically made to stand for an earlier epoch characterized by salvage and thrift. In what amounts to a moral and moralizing sociology, contemporary societies are then defined by their throwaway habits. Such accounts appear with regularity in the mainstream and popular media, and they are mobilized by the waste activist and sustainability literature (see Chapter 1). It is fair to say that they are very much the prevailing orthodoxy. For a useful summary, see: Hellman and Luedicke (2018). A rather different interpretation is to be found within a body of work in the critical social sciences that highlights the importance of a more serious sociological and historical engagement with these arguments. Rooted in evidence, it shows that contemporary societies are no more or less wasteful than their historical counterparts and that people continue to save and discard in equal measure. See: Gregson et al (2007a); Evans (2012); O'Brien (2013). The implication of these arguments, which I explore further in this chapter, is that to focus on consumers' alleged profligacy and wastefulness is to treat the symptom not the cause.

[8] Chiripanhura and Wolf (2019).
[9] Chiripanhura and Wolf (2019).
[10] Chiripanhura and Wolf (2019).
[11] There is a small but growing academic literature on the growth of Chinese consumption and Chinese consumer cultures. See, for example: Davis (ed 2000); Gerth (2003, 2010); Davis (2006); Hanser (2010); Tsang (2014); Yo (2014); Zhao (2014); Chrétien-Ichikawa (2015); Zhang (2017); Joy et al (2020). For a useful accessible summary, see: Hewitt (2007).
[12] Financial Times – Asian millennials ditch stigma for buying second-hand (www.ft.com); Financial Times – Second-hand market comes in from the cold in China (www.ft.com).
[13] McKinsey (2019).
[14] McKinsey (2019).
[15] See, for example: Milanovic (2016).
[16] Packard's text is seen as the first to make the connection between materialism, overconsumption and waste generation. Its critique of post-war American consumption as wanton, profligate consumerism continues to be highly influential and functions as the bedrock for arguments about the throwaway society.
[17] There is a considerable sociological literature that explores domestic technologies and their relationship to the gender division of labour – or, who does what household work, how much of such work is being done and its relationship to paid work. See Chapter 1, note 25.
[18] Wiens (2018).
[19] For a useful timeline, see fantasticfridges.com.
[20] The Diderot effect was first described by the French Enlightenment philosopher Denis Diderot in the essay 'Regrets on passing with my old dressing gown'. In this, he recounts the effects of being gifted a new red dressing gown – the result of which is to get rid of many of his other possessions: a straw chair, a desk, prints, and so on. New possessions, through their difference, can set in train the jettisoning of existing possessions and their

replacement with things that are more in keeping with things in the style of the new. The term 'the Diderot effect' was coined by Grant McCracken (1988), where he refers to the complementarity between consumer goods of a certain category, notably clothing and furniture and furnishings, and their relation to consumer identities. New possessions can either accord with how we see ourselves in our things, or they can be disruptive – as, for example, when clothing purchases relate to a new stylistic look. Typically, this is when jettisoning starts to occur, as an attempt to restore a coherent sense of who we are, as expressed in things. The Diderot effect doesn't just apply to the classic consumer goods. Major changes in living arrangements, such as moving house, are key precipitates for the Diderot effect, with 'old furniture' that accorded with the interior aesthetic of the former home often seeming 'out of place' in the new environment and hence being subjected to the same pressures.

21 Appliances that last (Which?, January 2021, p 25).

22 The most significant research on TV consumption is from the late 1980s and early 1990s. Located within a cultural studies and often explicitly feminist perspective, this research focused on what was going on when people watched TV. It was therefore at the heart of a core mission of cultural studies, which was to take popular culture, and popular media, seriously and give it the due academic attention that it merited. This research emphasized audiences and the domestic context of viewing, rather than the production of media; how TV watching was embedded in everyday life; and how TV watching constituted families. The key texts remain Morley (1986); Spigal (1992); Silverstone and Hirsch (eds 1992); Silverstone (1994).

23 This trend, which has run alongside the expansion in media consumption on the move, has been visible across multiple domains for some time. It first emerged in relation to food and alcohol consumption ('eating in' and drinking at home), and in gaming. More recently, there has been the growth in the home entertainment systems market. These trends were accelerated by the COVID-19 pandemic, in which the widespread adoption of national lockdowns led to enhanced consumption of streamed media. The market for smart TVs, for example, is estimated to be growing at 21 per cent per annum (www.grandviewresearch.com).

24 Wolfe (2008).

25 For a cultural history of Du Pont and its influence on American women's wardrobes, see: O'Connor (2011).

26 Bloomberg – China's next problem is recycling 26mt of discarded clothes (www.bloomberg.com).

27 On repair generally, see: Graham and Thrift (2007), who show its significance for urban infrastructure and modern life. One only has to think of the ongoing programmes of network repair and maintenance that apply to transport networks, utilities networks and communications networks across all cities in the global North to see that repair and maintenance are essential activities for the maintenance of economic, social and cultural life. See too: Persaud et al (2019).

28 There are exceptions to this (global North) rule, notably the mass programmes of 'slum clearance' and social housing renewal that have characterized periods of post-war European history. In the UK, this resulted in the widespread demolition of 19th century back-to-back terraced housing and, more recently, of their modernist replacements, or 'homes in the sky'. Another post-war exception has been Japan, where the dominant pattern has been for homes to be demolished every 20–30 years – see: Berg (2017). As in so many other areas, the climate emergency poses the biggest threat to the housing repair model. Retro-fitting insulation to combat building heat loss is one thing, but what is becoming increasingly apparent is that housing stock adaptation requires the wholesale reconfiguration of the infrastructures that heat, cool and connect homes. In

such circumstances, 'scrap and rebuild' is often the most cost-effective technical solution, but in countries where homes are assets, and where democracy is literally grounded in property owning, this will not be a political option.

29 Shove et al (2012).
30 Warde (2005).
31 See Southerton and Shove (2000); Shove (2003).
32 Gregson et al (2009).
33 Southerton (2013).
34 It should be noted, though, that even with bicycles, there is manufactured obsolescence. Key technological changes render replacement parts increasingly difficult to source, making it harder to effect repairs, eventually forcing the purchase of a new bike.
35 BBC (2019) – *Secrets of the Museum*.
36 An open question raised by new research on contemporary thrift is the extent to which repairing clothing and/or domestic furnishings is making a return. While there has been renewed interest in domestic craft, particularly among younger women (making things through knitting and sewing garments), the allied practice of repair is seemingly less widely practised, and potentially even polarized into those who enthusiastically engage in all manner of repair activities and those who don't. See: Holmes (2019).
37 The turn to community repair activities, through repair cafés and volunteer-led events, is an example of this, which seeks not only to establish the right to repair goods, but the skills and competences necessary to make effective repairs. See: The Restart Project (www.therestartproject.org).

Chapter 4

1 The fundamental idea behind the promotion of the sharing economy is that we will consume less collectively by sharing more. For key contributions, see: Belk (2014); Martin (2016).
2 Whether it will remain so is a moot point, and inevitably speculative. My own thoughts on this are that as capitalist economies become increasingly asset based (Piketty, 2014), then ownership models are here to stay, particularly with key goods such as housing, the ownership of which drives the acquisition of so many consumer goods.
3 The literature on regimes of value traces back to Arjun Appardurai's *The Social Life of Things*, whose broad definition I use here. The other key contributions that I draw on are: Igor Kopytoff's chapter in the same book, which first drew attention to the biography of things, and Fred Myers' *The Empire of Things*, which noted that regimes of value co-exist and interrelate, meaning that the value of an object is open to slippage, depending on which regime of value it is located within or which regimes it is moving between.
4 Ramsay (2009).
5 A yearling is a one-year-old horse.
6 This was demonstrated most clearly in a research project that examined UK household discard (Gregson et al, 2007a). In the course of two full years of investigation of the objects people got rid of from their homes, only 29 per cent of discards were routed directly to the waste stream, whereas 60 per cent were either given away to charity, friends and family, or sold.
7 Activities in the social economy coalesce around the social inclusion agenda. Many are targeted towards alleviating poverty and inequality and/or focus on employment options aimed at helping individuals reintegrate into wider society (for example, young offenders, former prisoners, service veterans). Others group around disability rights, access and mobility. While they can involve partnerships with capitalist firms (big as well as small and medium-sized enterprises), social economy interventions are typically third-sector led and heavily dependent upon voluntary labour.

8   Stallybrass (1998).
9   eBay traces back to 1995. It was launched in the US and concentrated first on the exchange of collectibles among hobbyists. In 2002, eBay acquired PayPal and by 2008 had expanded worldwide, with hundreds of millions of users. In 2020, an estimated 60 million items of used clothing and homeware alone were exchanged via eBay in the UK (Butler, 2021b). Gumtree (which was acquired by eBay in 2005) and Preloved are sites for classified advertisements. Gumtree's origins – as the name suggests – are in community exchanges between Australians and New Zealanders living in London. Now it encompasses communities and regions across all of the UK. Initially, Preloved specialized in secondhand and vintage clothing and homeware. It now includes over 500 categories of goods, including horses and pets. In addition to these general platforms, there are niche used goods platforms targeting particular markets. The most significant of these are Depop and Vinted, which are fashion trading sites. Vinted has 37 million users globally, including 1.2 million in the UK (Butler, 2021b).
10 A related development is the expansion of mainstream retailers into the used goods market in the UK. In 2021, Asda announced that it was testing out the sale of vintage secondhand clothing in 50 of its supermarkets. Both John Lewis and Ikea are testing schemes to sell used furniture, while the online retailer ASOS has seen vintage sales increase by 92 per cent – Butler (2021a, 2021b); BBC – Ikea starts buy-back scheme – www.bbc.co.uk.
11 Mauss (1925); Malinowski (1922). The subsequent literature is extensive. For key contributions, see: Gregory (1982); Parry (1986); Appadurai (ed 1986); Strathern (1988); Thomas (1991); Weiner (1992); Graeber (2011). For a short summary, see: Miller (2001).
12 Weiner (1992).
13 Ongka's Big Moka – Dir. Charlie Nairn. www.imdb.com.
14 Gregson et al (2000).
15 Holmes (2018).
16 Makov et al (2020).
17 In a quantitative survey of car boot sales in North-east England in 2012, this was the largest category of purchasing with such goods being acquired by 29 per cent of buyers – Gregson et al (2013).
18 Clarke (2000).
19 Being much less visible than any form of sale, the part of the hand-me-down-around economy discussed here is relatively unconsidered in the academic literature. The nearest example is Clarke's description of the thrice monthly packages received by 'Jane' from her US-based sister-in-law (2000 pp 86–7), but in this instance, and unlike the activities discussed here, the gifts are clearly not intended for return to their sender.
20 Gregson et al (2007b).
21 Recent surveys indicate that 74 per cent of adults living in the UK donated in the past 12 months to charity shops, with women, older people and religious people being more likely to donate than younger – Harrison Evans (2016). In part, this reflects the easy access to this conduit; in 2016, there were 10.8k charity shops in the UK, with most of these being on high streets.
22 Harrison Evans (2016).
23 Gregson and Crewe (2003).
24 Gregson and Crewe (2003).
25 Key contributions include: Nelson et al (2007); Carfagna et al (2014); Aptekar (2016); Eden (2017).
26 Aptekar (2016).
27 A related point is the connection between acquisition through some of these conduits and disposability. Cheap goods lead frequently to buying more but also encourage throwing things out, in much the same way as happens with fast fashion (Chapter 2). So it is that

things bought, or acquired, on impulse in these sites get taken home, but then, if they turn out to be unsuitable or 'bad buys', they rapidly get binned, precisely because they were so cheap in the first place – Gregson et al (2013). In such a way, some of the sites discussed here can actually accelerate the passage of things towards rubbish value, even while they appear to be extending the social life of things.

28 Johnson (1985); Tebbutt (1984).
29 Roberts and Zulfiqar (2019).
30 At the most extreme, 15 were recorded in one street in Liverpool alone – McMahon (2018).
31 McMahon (2018).
32 Boggan (2006); Rawlinson (2013).
33 The National Pawnbrokers Association – www.thenpa.com.
34 Roberts and Zulfiqar (2019).
35 Clarke (1998).
36 Gregson and Crewe (1997); Gregson et al (1997); Gregson and Crewe (1998).
37 Herrman and Soiffer (1984).
38 Gregson and Crewe (2003).
39 Kellett (2014).
40 For an exploration of this in relation to food waste caddies, see: Metcalfe et al (2012).
41 Chappells and Shove (1999).
42 Kellett (2014).
43 Gregson et al (2007a).
44 That waste collection services are paid for via local taxation undoubtedly helps to legitimate its use in this way too.
45 Gregson et al (2007b).
46 This charity is vertically integrated. Charities that lack this kind of sorting facility pay the price for that in the prices that they receive for unsold clothing. Current textile prices at the time of writing (Q1 2021) give an indication of the difference, with the price per tonne from textile banks being £50–125, from charity shops £200–300 and from sorting plants £370–430 – (www.letsrecycle.co.uk). Falling prices from textile recyclers have led to charities exploring different sales strategies, with several resorting to deep discounting and rapid turnover to shift stock – a move that is often marketed as illustrative of reuse over recycling (Harrison Evans, 2016). Given the connection between cheap goods and disposability (Chapter 3), it is unclear just how much actual reuse occurs with such goods.

Chapter 5

1 There is a considerable academic literature in the social sciences that focuses on scavengers and scavenging in the garbage dumps of the global South. Mostly this comes from a development and/or anthropological perspective. See: Medina (2007) for an overview. Medina estimates that around 2 per cent of the urban population of developing countries (or roughly 64 million) were then dependent on scavenging for their livelihoods. Examples of papers include: Birkbeck (1978); Furedy (1984); de Kock (1987); Meyer (1987); Huysman (1994); Tevere (1994); Beall (1997); Agunwamba (2003); Hayami et al (2006); Moore (2008); Oteng-Arabia (2012); O' Hare (2019). For a perspective from within the waste management literature, see: Wilson et al (2006); Wilson et al (2012).
2 The best example is *Our Mutual Friend*, in which Dickens explores the connections between dirt, dust and wealth in England's then rampant industrial capitalism.
3 An excellent illustration of this is Gill (2009), which focuses on plastic recycling in Delhi.
4 See: Bertoncelo and Bredeloup (2007); Mathews (2011); Mathews et al (2012); Mathews and Yang (2012); Lee (2014). In a similar vein, business journalist Adam Minter has shown

how Chinese scrap traders take advantage of the same short-stay visas to source various categories of scrap in the US – Minter (2013).

5. Mathews and colleagues' work has highlighted the importance of buildings like Chungking Mansions in Hong Kong as a hub for African traders doing business in China. Kenyan, Tanzanian, Nigerian and Ghanaian traders all come into Hong Kong on 90-day visas, using this as a base to place orders in mainland China. Those orders encompass mobile phones, computers, clothing, electronic goods, building materials, furniture and consumer goods from cigarette lighters to sunglasses, and are shipped by container, or in luggage, back to the main African free ports. See also: Neuwirth (2011).

6. In short, containers need to get back east to get filled up again with more export loads of consumer goods, and a reverse trade filled what would otherwise have been empty containers that nonetheless still cost to ship. For global shipping companies, the trade in discard made the back-run pay.

7. Key reports include: BAN (2002, 2005); Greenpeace (2021). For the equivalent academic literature, see: Clapp (2001); Pellow (2008).

8. Alternatively, it can be the result of straight (illegal) dumping. This typically occurs only when firms are paid to discard problematic (usually toxic) materials (see the Trafigura case under subheading 'A brief (and critical) note on toxic colonialism') and/or when imported materials turn out to be substandard and are rejected by potential processors and then sold on to third parties, who take the money and then dump the wastes.

9. For this account, I draw heavily on Blas and Farchy (2021, pp 233–8).

10. Alexander and Reno (eds 2012); Minter (2013).

11. Simon and Brooks (2012).

12. Hu et al (2020).

13. Minter (2020).

14. Minter (2020).

15. Gregson and Crang (2019).

16. For an indication of how 'foreign garbage' ends up in China, see: R v Biffa Waste Services Ltd – www.casemine.com; Creech (2020); Taylor (2021).

17. Creech (2021).

18. Minter (2020).

19. Hansen and Le Zotte (2019).

20. The sites of primary resource extraction are typically concentrated and reflect the economics of extraction more than the physical location of resources. It is only when sufficient resource of a particular grade occurs in proximity that mining is deemed economic, and mines (or wells) are established. Recovered resources, by definition, need to be collected to be recovered. And therein lies a major economic challenge, for these resources can be literally anywhere, making the costs of their collection high.

21. Thompson (1963).

22. Shoddy is the technical term used for fibre that has been derived from pulling softer, usually unmilled cloth; mungo is derived from harder, mostly milled cloth and is a finer product. Often the term 'shoddy' is used generically to encompass all recovered textile fibre. For an exploration of shoddy as an historical product, see: Shell (2020).

23. For useful summaries, see: Shell (2014); Shoddy and Mungo – kirkleescousins.co.uk.

24. The term 'virgin' fibre was adopted specifically to foreground shoddy's material deficiencies and lies behind the subsequent use of the term 'shoddy' to denote low-grade, poor-quality work or products (Shell, 2020).

25. Additionally, and in a mesh of public health concerns with cultural taboo, questions were raised over the use of the uniforms and overcoats of dead soldiers for stuffing mattresses.

26. Quoted in Shell (2014).

NOTES

27 Jenkins estimates that in 1880, 40 per cent of the clean wool weight used by the British woollen industry was made up of recovered wool; by 1913, that was up to 60 per cent and in Yorkshire it comprised 75 per cent. He estimates that the cost advantages of a one-third new/two-thirds recovered blend would be around 13 per cent and argues that it is shoddy that allowed Britain to compete successfully on price in emerging global woollen markets, accounting for an estimated 46 per cent of global trade by 1913. In addition, using recovered wool allowed manufacturers to make changes to blends in response to changes in the relative price of raw and recovered wool. See: Jenkins (1990); Jenkins and Malin (1990).

28 American manufacturer quoted in Jenkins and Malin (1990 p 76).

29 Jenkins and Malin (1990).

30 Bhangat P (nd) Panipat woollen industry, Asia's blanket business hub in Asia – www.filmingindo.com.

31 Prato's success as the major European woollen centre of the latter half of the 20th century was also down to its use of recovered fibre in its wool blends – Jenkin and Malin (1990).

32 See the documentary 'short': *Unravel* – Meghna Gupta (2012).

33 Figure 1: Illustrating global flows of used clothing – Gregson and Crang (2015).

34 The story of Asia's biggest textile recycling hub – dwijproducts.com.

35 Norris (2012, p 397).

36 Panipat, the global centre for recycling textiles is fading – economist.com; In Panipat, the world's 'castoff' capital' business hangs by a thread – www.hindustantimes.com.

37 In Panipat, the world's 'castoff' capital' business hangs by a thread – www.hindustantimes.com.

38 Minter (2018).

39 *Unravel* – Meghna Gupta (2012).

40 Morley et al (2009) estimate that only 20 per cent of discarded used clothing (by tonnage) in the UK ends up being reused in the UK (Table 5.2 p 12). Outside of social networks, the primary means to this is charity shops (or, goodwill or thrift stores – see Chapter 4). The remainder (comprising textile bank and return schemes and any unsold charity shop stock) heads directly to textile sorting factories. So, these factories are handling an estimated 80 per cent of the UK's discarded clothing by tonnage.

41 There is an old sociological literature that provides key insights into assembly line working; see the classics: Beynon (1973); Cavendish (1982).

42 Botticello (2012).

43 I draw here extensively on Botticello (2012). See too: Gregson et al (2016).

44 Thorpe (2016).

45 The pioneering work on West African clothing reuse markets is Karen Hansen's study of Zambia, *Salaula* (2000).

46 Thorpe (2016) provides an account of one such plant, which sorts 100 tonnes/day. The general point is that cheaper labour in Eastern Europe not only allows for extracting more value from clothing discard, but also allows sorting factories located there to capture a share of the global value chain. Effectively, they have become intermediaries between Northern Europe and the markets of the global South.

47 Rivoli (2005).

48 Crang et al (2013).

49 Abimbola (2012).

50 This pattern is not just about the importance of the UK as a global exporter of used clothing; it has also been aided by recent visa regimes.

51 The EU End-of-life Vehicles Directive came into European Law in 2000 and is widely regarded as the first example of legislation to illustrate the principles of producer responsibility. Aside from prohibiting the use of particular substances in car manufacture

52. (lead, mercury, cadmium and hexavalent chromium), it has led to the development of 'take-back' schemes and sets reuse and recycling targets for end-of-life cars – currently 95 per cent recovery and 85 per cent recycling, by weight. See: End of life vehicles – https://environment.ec.europa.eu/topics/waste-and-recycling/end-life-vehicles_en
52. Gregson et al (2013).
53. Merchant ships are registered under 'flags of convenience' and it is commonplace for ships to be registered under multiple flags as they are bought and sold during their working lives. These flags work much like tax havens; as well as being light in regulatory compliance compared with national registries (which are aligned with the law in their associated territorial jurisdictions), they render lines of ship ownership opaque. See: von Fossen (2016). In this way, it is virtually impossible to connect a merchant ship that ends up on a beach in South Asia with a particular owner.
54. This is the Layal Yard at Aliağa – see: Warship HMS Invincible broken up in Turkish port (www.bbc.co.uk). Subsequently, when the bottom fell out of the cruise ship market on account of the COVID-19 pandemic in 2020, the Aliağa yard saw a boom in cruise ship scrapping – Narishkin et al (2021).
55. Gregson et al (2013).
56. Standard practice would see this material being sent for either landfill or incineration, or potentially to what is known in the waste management business as a 'dirty MRF' for further sorting, but it should be noted that all these routes incur costs.
57. Gregson et al (2015).
58. Gill's (2009) study of plastics recycling in Delhi identifies two main grades structuring the market: Grade 1 (primary plastic, produced from virgin polymers) and Grade 2 (mixed materials, including recycled resins and polymers). In this way, we can see how recovered materials are regarded as of inferior quality to their primary counterparts – a material difference that is reflected in prices. The grades are also crosscut by contamination levels.
59. World Economic Forum (2016).
60. The British Plastics Foundation notes partnerships between LyondellBasell and Suez, pilot schemes involving BASF and Remondi, and a memorandum of understanding between SABIC and Plastic Energy, with project partners that include Unilever and Tupperware – https://www.bpf.co.uk/plastipedia/chemical-recycling-101.aspx#_Toc31632543
61. Jennings (2017).
62. Laird (2016).
63. Commercialization of plastic waste derived fuel for generating electricity – gtr.ukri.org.
64. White (2018).
65. White (2018).
66. Figures in the public domain suggest a saving of £90/tonne on gate fees versus selling at £300/tonne – White (2018).
67. See Wei and Zimmermann (2017).
68. Carrington (2021) Scientists convert used plastic bottles into vanilla flavouring – www.guardian.co.uk; Sadler and Wallace (2021).
69. See: Bomgardner (2016).
70. Bensaude-Vincent and Stengers (1996).
71. Verschoor et al (2021).
72. The circular economy argument has strong affinities with industrial ecology, the case for which was first sketched out by Frosch and Gallopoulous (1989). However, one major difference between the two is the ecological metaphor. In seeking to think of manufacturing through ecological principles, proponents of industrial ecology were drawn to push for the merits of linkages between proximate firms, and therefore saw this as a distinctly spatial policy. The problem for industrial ecology turned out to be that precious few instances of such arrangements developed organically in capitalist economies (the

example of Kalundborg in Denmark is repeatedly recited), while policies to promote such arrangements were conspicuously unsuccessful – in contrast to the situation in China (Gregson et al 2012). By comparison, circular economy arguments are aspatial; the arguments are made with no reference to geographical space. Rather, they are presented as environmentally desirable and as a business and/or technical innovation opportunity.

73 Desrochers (2002); Desrochers (2008).
74 In 1928, the chemical engineer John Kershaw wrote in his book *The Recovery and Use of Industrial and Other Wastes*, 'Dirt from the philosopher's standpoint is simply matter in the wrong place, and industrial waste may be regarded similarly as useful material produced or dumped in places where it is not required. When transported to the right spot an industrial waste will often form the raw material for some secondary industry or manufacture' – quoted in Desrochers and Lam (2007). In that quote, we see how the elision between dirt and waste as matter out of place is made; see Chapter 2.
75 Gregson et al (2012). See too: Lepawsky (2018).
76 Gregson et al (2010).
77 Blas and Farchy (2021).

Chapter 6

1 There are multiple versions of the Zero Waste argument. For waste activists, this is often interpreted literally – a situation that is a material impossibility. However, when regulators and industry representatives use the term, they are primarily alluding to a world in which landfills are made increasingly redundant by the turn to greater resource efficiency, through enhanced resource recovery, recycling and reuse. In other words, Zero Waste is intricately bound up with ideas about circularity.
2 There is a considerable academic literature charting how finance has become the most significant part of contemporary capitalist economies. For a useful review, see: van der Zwan (2014). Van der Zwan identifies three strands to financialized capitalism: financialization as a regime of accumulation; the rise of shareholder value as the guiding principle of economic life; and the financialization of everyday life, which entails not only turning citizen consumers into investors but also the tendrils of finance being extended into all walks of ordinary life. As will become clear through this chapter, when discard becomes captured as waste, we are talking primarily about an instance of the third tendency, although shareholder value is also never far away. More recently, it has been recognized that there is a fourth component to financialization: assetization (see: Birch and Muniesa, 2020; Langley, 2021). Birch and Muniesa contend that assetization has become the primary basis of contemporary capitalism, overtaking the role of commodification.
3 An asset is defined as anything that can be controlled, traded and capitalized as a revenue stream (Birch and Muniesa, 2020). While commodities are bought and sold, the key difference that marks out an asset is the capacity for it to become the basis for extracting economic rent, as a revenue stream, into the future.
4 Valuation relies on net present value – a means of calculating the present worth of future investment earnings. Net present value relies on discounting; a raft of assumptions is made about the future value of money, based on assumptions about inflation, risks and other uncertainties. In such a way, investors and fund managers make comparisons between various possible investments and their relative anticipated earnings. See: Dorganova (2018).
5 Currently, in the global North, even countries with an established recycling collection infrastructure send over 50 per cent by weight of household discard to the residual waste stream.
6 Strasser (1999).
7 Hoornweg and Bhada-Tata (2012); Kaza et al (2018).

8. To qualify as 'R1', energy-from-waste plants have to generate an energy output that is 65 per cent (or more) of the calorific value of the waste input. For a discussion, see: Behrsin (2019); for the internal trade, Behrsin and de Rosa (2020).
9. Gille (2007).
10. Perhaps the most famous example of this is 19th-century Paris. Haussmann's programme of modernization works is widely regarded as the biggest urban renewal programme enacted in any major city in a period of peace. It involved the demolition of major parts of the medieval city; the construction of wide avenues, parks and squares; and the construction of a huge network of sewers along with fountains and aqueducts. A parallel programme of sewer improvement occurred in London under the oversight of Bazalgette. See: Gandy (1999); Cadbury (2002). On the connection in the 19th century between the sanitary approach and the emerging scientific understanding of bacteria and germs, see: Barnes (2006); Melosi (2008).
11. In some parts of the world, those responsibilities are shared by city-wide administrations; in others (England is an example), they may be both shared and divided, with different local government administrative areas having responsibility for collection and disposal. This creates antagonisms, for the interests of collection authorities do not necessarily coincide with those of disposal authorities. Most notable here is the difference in financial position. Collection authorities have to pay disposal authorities for waste disposal services.
12. Cooper (2010); Herbert (2007).
13. Cooper (2010 p 1036).
14. In the following section I draw heavily on Cooper (2010).
15. The global poster child for landfill restoration is Fresh Kills Park, Staten Island, New York. During its life as a landfill, Fresh Kills attracted much negative comment – for its odour, for the plastic that escaped its confines and for its scale. Often cited as the world's largest landfill, it was seen to represent 20th-century America's approach to consumption and to garbage – see: Melosi (2020). Following its closure in 2001, the 2,200-acre site has been transformed through a restoration project, which over the course of the past 20 years has covered over four mounds of trash, generating forest and scrubland and open grass fields, and encouraging the formation of tidal wetlands. To that have been added trails for hiking and cycling. See: *The Fresh Kills Story: From World's Largest Garbage Dump to a World-class Park* (2012); Sullivan (2020). Elsewhere in the US, other former landfills have been restored in ways that favour wilderness restoration over parkland amenity value – see, for example, Coyote Canyon CA. The Puente Hills restoration programme outside Los Angeles is scheduled to include both types of restoration practice.
16. Fresno's status as a Landmark site was the subject of much controversy. Already designated a Superfund site (a hazardous waste site in need of clean-up), many argued that it could not possibly be a heritage site. Others, taking their cues from the commemoration of 'difficult' heritage, argued the opposite; Fresno's commemoration was not just a means to signal the importance of landfill to 20th-century urban life in the US, it was also a suitable indictment of US environmental policy under the presidency of George W. Bush. See: Melosi (2002); Wright (2018).
17. Gutierrez and Webster (2012).
18. Nowhere is that dependency more clearly illustrated than by what has happened to New York City's waste post the closure of Fresh Kills. Staten Island is now home to a giant waste transfer station and associated railhead, from which waste is trained to landfills in the Carolinas. See *The Fresh Kills Story: From the World's Largest Garbage Dump to a World-class Park* (2012), where this shift in disposal practices is paraded as an example of 'clean and green' waste management. Shipping New York's waste to the Carolinas shows that the geographical displacement of waste in the US has shifted scale. Whereas previously it was proximate poorer boroughs that were the dumping ground for New York City's

NOTES

wastes, now it is poorer states that have assumed that role. The closure of landfills adjacent to other affluent US cities is likely to further exacerbate these tendencies.

[19] Cui et al (2020).
[20] Minter (2013).
[21] Landfill waste – www.ec.europa.eu.
[22] See, for example: Rootes (2006); Rootes and Leonard (2010).
[23] Gregson and Forman (2021). The alternative solutions were either to pay for landfill or short-term commercial contracts with other energy-from-waste operators.
[24] Japan, Denmark, Sweden and Switzerland are all examples of countries with high levels of energy from waste and relatively low levels of recycling (Cui et al 2020).
[25] There are numerous examples of Chinese cities that exemplify using industrial linkages of wastes and by-products to circumvent the generation of wastes. There are several studies of firm-based symbiotic relationships, along with other studies of successful eco parks and of closed-loop thinking in economic development. For examples, see: Zhu et al's (2007) work on the Guitang Group of sugar refiners and Shi et al's (2010) study of the Tianjin Economic-Technological Development Area.
[26] Song et al (2013).
[27] Zhang et al (2015).
[28] Zheng et al (2014); Zhao et al (2016).
[29] Cui et al (2020).
[30] Yang (2019); Allen (2019)
[31] An earlier programme of source segregation was introduced in 1998, but failed to take hold.
[32] Quite what interests are behind these 'waste collection services' is a moot point: although it is possible that collection is being seen by some as a business opportunity, one might also speculate that this is the informal sector mobilizing digital platforms to retain access to discard.
[33] Shapiro-Bengsten (2020).
[34] It is well recognized in the technical literature that China's organic waste fraction poses problems for incineration. Material characterization work has shown a moisture content that puts China's municipal solid waste on the margins of viability for incineration, and early Chinese-built energy-from-waste plants typically resorted to burning coal for this reason. Initially, imported technology, developed in relation to drier input materials, struggled to cope with large volumes of 'wet' feedstock, although subsequently it has been successfully modified. The wider question of why Chinese municipal solid waste is so comparatively wet points to a rather broader issue, which is the role of the informal sector in extracting value from what becomes the waste stream. Or, in other words, 'wet' is only 'wet' when much that is 'dry' is removed.
[35] Shapiro-Bengsten (2020).
[36] Schulz (2015); Inverardi-Ferri (2017).
[37] Tong and Tao (2015).
[38] Chi et al (2014).
[39] Gill (2009).
[40] Gill (2009 p 248).
[41] Gill (2009).
[42] Rejh (2021).
[43] Ministry of Housing and Urban Affairs nd.
[44] Chattopadhay et al (2009); Talyan et al (2008).
[45] Casiella (2019).
[46] Rejh (2021 p 186).
[47] Composting has been consistently favoured in India – an indication of the continued importance of agriculture in the economy. Sharhooly et al (2008) document the extension

of composting to bigger cities under the 5th Five-Year Plan, with 150–300 tonnes/day facilities established in Bangalore, Baroda, Mumbai, Kolkata, Delhi, Jaipur, Kanpur and Indore. Nonetheless, the plants faced operational challenges and problems of marketing output (see: Chattopadhay et al 2009; Talyan et al 2008). The 2005–2021 Plan continued to favour composting, yet operational challenges clearly remain at scale.

48 Talyan et al (2008); Nixon et al (2017).

49 Delhi (2016 – 1200 tonnes/day); Guntur (1200 tonnes/day), Tirupati (450 tonnes/day) and Visakhapatnam (1200 tonnes/day) due 2021; and Tahkhand (1200 tonnes/day) due 2023 – www.martingmbh.de.

50 For discussion of the challenges of PPPs in India, see: Nixon et al (2017); Dolla and Laisham (2021).

51 Veolia to build 1.6m TPA waste to energy plant in Mexico City – www.waste-management-world.com.

52 See GAIA – the Global Alliance for Incinerator Alternatives (www.no-burn.org). This is an alliance of over 800 grassroots environmental justice groups spanning 90 countries. Many of the alliance campaign under the Zero Waste banner.

53 That focus can be traced to the earlier generation of incinerators, which burned wastes at lower temperatures than modern incinerators, and when there were also less stringent controls on permissible emissions. It was given added impetus by epidemiological research of the time that posited associations between incinerators and clusters of above average rates of cancer and leukaemia, most especially in children.

54 Typically, a municipal energy-from-waste contract would provide a guaranteed 50 per cent of the annual capacity of a plant, with the operating firm needing to source additional commercial contracts to increase plant throughput, and to increase revenue.

55 Perhaps the best example of this is Copenhagen's Copenhill plant, opened in 2019, which is topped by a dry ski slope, along with picnic facilities and tree-lined hiking trails, and which comes with its very own climbing wall. Designed by Bjarke Ingels Group, this is described as an example of 'hedonistic sustainability', by which it is meant that the plant is not just good for the environment but that it is good, too, for human well-being. See Slavin (2016); Crook (2019). Energy-from-waste plants are moving away from being nondescript, anonymous 'shed structures' to designs to be embraced by industrial architecture – see Hartman (2010). In China, a similar trajectory is underway, with plants being designed with visitor walkways that give spectacular city-wide vistas. Using architecture to attend to public distrust and controversy is not a new tactic, but when applied to energy-from-waste it signals a new, unapologetic stance to the technology and its place in cities. Far from hidden from view, this is waste treatment reclaiming its space in the city through the kinds of claims to perspective that have more typically accompanied the world's tallest buildings and their viewing platforms.

56 I draw here on the extensive information available via www.martingmbh.de. Note particularly, the detailed company history, technology-specific documents and the data set: Thermal waste treatment facilities using Martin technologies.

57 The scale of Chinese infrastructural investment in Africa is impossible to overstate. Since 2011, China has become the largest financier of African infrastructure projects, responsible for one in five projects and one in three construction projects – particularly in relation to transport, shipping and ports, and energy and power. See: Marals and Labushange (2019, Figs. 3–5). Chinese finance has entailed low interest rates but comes with resource-backed collateral, leading many to see this as a new form of resource-based imperialism in Africa. Infrastructure investment to date has been focused primarily on railways, ports and some (coal-fired) power stations, but such is the rate of urbanization in Africa that McKinsey estimates that by 2025, there will be over 100 African cities with populations

of over 1 million. It is this emerging market in municipal waste that the major players in the energy-from-waste market would appear to have their eyes on.

58 www.babcock.com.
59 Wilkie (2019).
60 NS Energy – Shenzhen East waste-to-energy plant – www.nsenergybusiness.com.
61 One of the key challenges facing the energy-from-waste sector in Europe and North America is that, as more and more material is either kept out of the residual waste stream or recovered prior to its treatment as residual waste, so the calorific value realized from combusting residual waste is reducing. Capturing material for recycling has an effect not just on the quantity of material available for residual processing; it also influences the material composition of the feedstock entering residual waste processing. That, I suggest, will have economic effects.
62 It is against this backdrop that the turn to plastics recycling needs to be seen (see Chapter 5). Up until very recently, the absence of plastic recycling markets meant that most plastic waste ended up in energy-from-waste plants. Removing plastic from residual waste feedstock will have an effect on the calorific value being generated from burning residual waste.
63 See note 10.
64 Barjot (2011) highlights the significance of the French 'concession' model (through which public bodies granted private companies exclusive rights to supply a municipal service) to the success of French companies (including Veolia and Suez) in the emerging international market in municipal services post 1990.
65 Veolia (2019) Annual financial report 2018 (Appendix 7, p 38) – www.veolia.com. Hazardous waste accounted for 20 per cent of waste revenues; energy from waste 19 per cent. By comparison, recycling services generated only 8 per cent – an indication of the collapse of prices consequent upon the China ban, but also of the volatility of revenue from recycling operations. The latest figures show a large growth in revenue from hazardous waste management, largely as a response to the COVID-19 pandemic and consequent huge uplift in medical grade PPE waste – Veolia (2021b) Integrated report.
66 Veolia – Half-yearly annual financial report 2021a – www.veolia.com.
67 Veolia (2021b) Integrated Report.
68 Veolia (2021c) Presentation: Creating the world champion of the ecological market – www.veolia.com.
69 In comparison to much of continental Europe, the UK was a late entrant to the energy-from-waste market. This was on account of its historic reliance on landfill – see previous section. But it also rested on strong public opposition to incineration – see note 53. It was not until 1999/2000, in response to the EU Landfill Directive, that the country had to engage at scale with energy-from-waste solutions. We see in the period post 2000 the emergence of an important new market in residual waste in Europe, and the opportunity therefore for all the major firms (and new entrants) to get a slice of that market. See Gregson and Forman (2021).
70 Tolvik Consulting (2021 p 6).
71 In 2022, following an inquiry into competition concerns by the Competition and Markets Authority, Veolia sold Suez UK to Macquarie Group for €2.4 billion.
72 Appelbaum and Batt (2014, 2017).
73 Guy Hands firm nets £1bn profit from waste disposal – www.thetimes.co.uk.
74 Biffa began life at the start of the 20th century collecting ash, dust and clinker in the London area, moved on to shifting sand and gravel and then into the industrial waste market, before it was acquired by an industrial services group and then sold to Severn Trent (a recently privatized water company) in 1991. The acquisition of UK Waste in 2000 and then Hales Waste in 2003 saw a concerted move into the emerging

waste-to-resource market, and in 2006, Biffa demerged from Severn Trent and became a publicly listed company.

75 Global Infrastructure Partners is an investment fund manager specializing in infrastructure assets, including energy and transport, and – at the time of Biffa's purchase – water and waste. For the attempted 'fire sale', see: Ebrahini (2012) Biffa owners eye quick fire sale after £1.2bn writedown – www.telegraph.co.uk; Mackie (2012). For the recapitalization, see: Eminton (2013) Biffa completes £1bn recapitalization. The package with hedge fund owners (Angelo Gordon & Co, Avenue Capital Group, Babson Capital Europa Ltd and Sankaty Advisors) involved a debt write-down of 55 per cent (from >£1.1 billion to £520 million) and an injection of £75 million. By 2015, Biffa had returned to a sounder financial position – Biffa eyes more acquisitions as it returns to growth – www.letsrecycle.co.uk.

76 In light of this experience, it is significant to note that while it remains an important player in the UK's collections and recycling market, Biffa (notwithstanding having had aspirations in this direction) has a limited presence in the UK energy-from-waste market. That is possibly not unrelated to its financial travails post 2008, which was the point at which the UK municipal waste market was at peak procurement.

77 KKR enters UK's energy from waste sector – www.infrastructureinvestor.com.

78 Viridor 'to sell waste and recycling subsidiary' – www.mrw.co.uk; Sale of Viridor assets to Biffa remains 'on track' – www.letsrecycle.co.uk.

79 EQT in $5.3bn take-private deal for waste-to-energy leader Covanta – www.investableuniverse.com.

80 Park (2021).

81 The Enron case was the first to show how SPVs can hide debt and toxic assets from potential investors and creditors. At the time (2001/2), it was the biggest corporate bankruptcy to hit the financial world and was only surpassed in the global financial crisis. See: Segal (2021) for an account.

82 Allen and Pryke (2013); Loftus and March (2016); Pryke and Allen (2019).

83 Spotlight: China's waste-to-energy industry – www.fitchratings.com.

Chapter 7

1 The restaurants were introduced to respond to anticipated food scarcity and shortages. They existed for the period 1940–1947, continuing beyond the war into the early period of post-war austerity. Most towns of over 50,000 people had them, and some smaller towns. They provided a choice between five meat dishes, five vegetable dishes and a dessert, with a two-course meal costing 9d (roughly £1 today). For a summary, see: Atkins (2011).

2 See: Zuboff (2019), particularly the sections on Google and extracting the behavioural surplus.

3 See: Bogner et al (2007). The waste management sector's contribution to GHGs was then estimated at ~3 per cent, meaning that the sector as a whole was not seen as a major challenge; landfill was, however. Moreover, in retrospect it is possible to see that intervening in relation to landfill was an easier task than the current challenge of reducing methane emissions from livestock production. Reducing consumers' meat consumption is a much harder policy goal than reducing landfill by introducing recycling habits.

4 Actual estimates for the sector are difficult to come by. They vary at the plant level, depending on the overall feedstock composition, and across a year. ^{14}C measurements of emissions from three plants located in Zurich put the biogenic:fossil fractions at roughly 50:50. See: Mohn et al (2008).

5 See: Pour et al (2018).

6 Astrup et al (2009).

7 For an overview, see: Ryan and Rothman (2022).

References

Abimbola O. (2012) The international trade in secondhand clothing: managing information asymmetry between West African and British traders, *Textile* 10: 184–99.

Abramovitz S. (2017) Epidemics (especially Ebola), *Annual Review of Anthropology* 46: 421–45.

Abramovitz S., Bardosh K., Leach M., Hewlitt B., Nichter M., Nguyen V-K. (2015) Social science intelligence in the global Ebola response, *The Lancet* 385(9965): 330.

Acemoglu D., Restrepo P. (2020) Robots and jobs: evidence from US labor markets, *Journal of Political Economy* 128(6): 21–44.

Afreen's Kitchen (2018) Pakistani menu ideas for Eid lunch/dinner – www.afreenskitchen.wordpress.com (6 June 2018).

Agunwamba J. (2003) Analysis of scavengers' activities and recycling in some cities in Nigeria, *Environmental Management* 32(1): 116–27.

Akhtar A., Sarmah A. (2018) Construction and demolition waste generation and properties of recycled concrete: a global perspective, *Journal of Cleaner Production* 186: 262–81.

Alexander C., Reno J. (eds 2012) *Economies of Recycling* (London: Zed Books).

Alexander C., Smaje C. (2008) Surplus retail food redistribution: an analysis of a third sector model, *Resources Conservation and Recycling* 52: 1290–98.

Alexandersen S., Kitching R., Mansley L., Donaldson A. (2003) Clinical and laboratory investigations of five outbreaks of foot and mouth disease during the 2001 epidemic in the UK, *Veterinary Record* 152: 489–96.

Allen J., Pryke. M (2013) Financialising household water: Thames Water, MEIF, and 'ring-fenced' politics, *Cambridge Journal of Regions, Economy and Society* 6(3): 419–39.

Allen K. (2019) Shanghai rubbish rules – www.bbc.co.uk (4 July 2019).

Appadurai A. (ed 1986) *The Social Life of Things* (Cambridge: Cambridge University Press).

Appelbaum E., Batt R. (2014) *Private Equity at Work* (New York: Russell Sage Foundation).

Appelbaum E., Batt R. (2017) How private equity firms are designed to earn big whilst risking little of their own – www.blogs.lse.ac.uk (23 January 2017).

Aptekar S. (2016) Gifts among strangers: the social organisation of Freecycle, *Social Problems* 63: 266–83.

Astrup T., Møller J., Fruergaard T. (2009) Incineration and co-combustion of waste: accounting of greenhouse gases and global warming contributions, *Waste Management and Research* 27: 789–99.

Atkins P. (2011) 'Communal feeding in war-time: British Restaurants 1940–47'. In Zweiniger-Bargielowska I., Duffett R., Drollard A. (eds) *Food and War in Twentieth Century Europe* (Farnham: Ashgate), pp 139–53.

Ault A. (2015) Ask Smithsonian: are cats domesticated? – www.smithsonianmag.com (30 April 2015).

Babcock & Wilcox – https://www.babcock.com/ (last accessed 13 December 2022).

BAN (2002) *Exporting Harm: the high tech trashing of Asia* (Seattle: Basel Action Network).

BAN (2005) *The Digital Dump: exporting reuse and abuse to Africa* (Seattle: Basel Action Network).

Barjot D. (2011) Public utilities and private initiative: the French concession model in historical perspective, *Business History* 53: 782–800

Barnard A. (2011) 'Waving the banana' at capitalism: political theatre and social movement strategy among New York's 'freegan' dumpster divers, *Ethnography* 12(4): 419–44.

Barnard A. (2016) *Freegans* (Minneapolis: University of Minnesota Press).

Barnes D. (2006) *The Great Stink of Paris and the nineteenth century struggle against filth and germs* (Baltimore: John Hopkins Press).

Bazerghi C., McKay F., Dunn M. (2016) The role of food banks in addressing food insecurity: a systematic review, *Journal Community Health* 41(4): 732–40.

BBC (2011) HMS Invincible broken up in Turkish port – https://www.bbc.co.uk/news/uk-england-13778654 (15 June 2011).

BBC (2019) Glastonbury 2019: Festival clean up underway – www.bbc.co.uk (1 July 2019).

BBC (2019) *Secrets of the Museum* (Blast! Films).

BBC (2021) Ikea starts buy-back scheme – www.bbc.co.uk (5 May 2021).

Beall J. (1997) Thoughts on poverty from a South Asian rubbish dump, *IDS Bulletin* 28(3): 73–90.

Beall J. (2006) Dealing with dirt and the disorder of development: managing rubbish in urban Pakistan, *Oxford Development Studies* 34: 81–97.

Behrsin I. (2019) Rendering renewable: technoscience and the political economy of waste to energy regulation in the EU, *Annals of the American Association of Geographers* 109: 1362–78.

Behrsin I., de Rosa S. (2020) Contaminant, commodity and fuel: a multi-sited study of waste's role in urban transformations from Italy to Austria, *International Journal of Urban and Regional Research* 44: 90–107.

Belk R. (1995) *Collecting in a Consumer Society* (London: Routledge).

Belk R. (2014) You are what you can access: sharing and collaborative consumption online, *Journal of Business Research* 67: 1595–600.

Bennett J. (2004) The force of things: steps toward an ecology of matter, *Political Theory* 32: 347–72.

Bensaude-Vincent B., Stengers I. (1996) *A History of Chemistry* (Cambridge MA: Harvard University Press).

Berg N. (2017) Raze, build, repeat: why Japan knocks down its houses – www.theguardian.com (16 November 2017).

Bergmans A., Kos D., Simmons P., Sundqvist G. (2015) The participatory turn in radioactive waste management: deliberation and the socio-technical divide, *Journal of Risk Research* 18: 347–63

Berman M. (1987) *All That is Solid Melts Into Air* (New York: Verso).

Bertoncelo B., Bredeloup S. (2007) The emergence of new African 'trading posts' in Hong Kong and Guangzhou, *China Perspectives* 2007/1.

Beynon H. (1973) *Working for Ford* (London: Allen Lane).

Bhangat P. [nd] Panipat woollen industry, Asia's blanket business hub in India – www.filmingindo.com.

Bickerstaff K. (2012) 'Because we've got history here': radioactive waste, community volunteerism and the haunting of environmental politics, *Environment and Planning A* 44: 2611–28.

Birch K., Muniesa F. (eds 2020) *Assetization* (Cambridge MA: MIT Press).

Birkbeck C. (1978) Self-employed proletarians in an informal factory: the case of Cali's garbage dump, *World Development* 6(9/10): 1173–85.

Bittman M., Rice J., Wajcman J. (2004) Appliances and their impact: the ownership of domestic technologies and time spent on household work, *British Journal of Sociology* 59: 59–77.

Blas J., Farchy J. (2021) *The World for Sale* (London: Cornerstone).

Bloom J. (2011) *American Wasteland* (Boston: De Capo Press).

Bloomberg (2020) China's next problem is recycling 26mt of discarded clothes – www.bloomberg.com (18 October 2020).

Blowers A. (2016) *The Legacy of Nuclear Power* (London: Routledge).

Bluestone B., Harrison B. (1982) *The de-industrialisation of America* (New York: Basic Books).

Boggan S. (2006) Pawn again – www.theguardian.co.uk (20 October 2006).

Bogner J., Abdelrafie A., Diaz C., Faaij A., Aao Q., Hashimoto S. et al (2007) 'Waste Management'. In Metz B., Davidson O., Bosch P., Dave R., and Meyer L. (eds) *Climate Change: Mitigation*. Contribution of Working Group III to the 4th Assessment Report of the Intergovernmental Panel of Climate Change (Cambridge: Cambridge University Press). Available at: https://www.ipcc.ch/site/assets/uploads/2018/02/ar4-wg3-chapter10-1.pdf (last accessed 5 February 2023).

Bolter H. (1996) *Inside Sellafield* (London: Quartet Books).

Bomgardner M. (2016) The problem with vanilla: after vowing to go natural food brands face a shortage of the favoured flavour – scientificamerican.com (14 September 2016).

Booth S., Whelan J. (2014) Hungry for change: the food banking industry in Australia, *British Food Journal* 116(9): 1392–404.

Bordo S. (1993) *Unbearable Weight* (Berkeley: University of California Press).

Botticello J. (2012) Between classification, objectification and perception: processing second hand clothing for recycling and reuse, *Textile* 10: 164–83.

Boulding K. (1966) 'The economics of the coming spaceship.' In Jarett H. (ed) *Environmental Quality in a Growing Economy* (Baltimore MD: Johns Hopkins University Press).

Brennan S. (2018) *Stamping American Memory* (Ann Arbor: University of Michigan Press).

British Plastics Foundation (2022) Chemical Recycling 101– https://www.bpf.co.uk/plastipedia/chemical-recycling-101.aspx#_Toc31632543 (last accessed 16 December 2022).

Butler (2021a) Asda to sell vintage clothes – www.theguardian.co.uk (28 April 2021).

Butler (2021b) Preloved – www.theguardian.co.uk (1 May 2021).

Butt W. (2019) Beyond the abject: caste and the organisation of work in Pakistan's waste economy, *International Labor and Working Class History* 95: 18–33.

Butt W. (2020) Waste intimacies: caste and the unevenness of life in urban Pakistan, *American Ethnologist* 47: 234–48.

Cadbury D. (2002) *Seven wonders of the industrial world* (London: Fourth Estate).

Calvignac C., Cochoy F. (2016) From 'market agencements' to 'vehicular agencies': insights from the quantitative observation of consumer logistics, *Consumption, Markets and Culture* 19(1): 133–47.

Cambridge Independent (2020) Concern over traffic and littering after re-opening of Newmarket Road McDonald's, Cambridge – www.cambridgeindependent.co.uk (13 June 2020).

Caplow T. (1982) Christmas gifts and kin networks, *American Sociological Review* 47: 383–92.

Carfagna L., Dubois E., Fitzmaurice M., Oulmette M., Schor J., Willis M. et al (2014) An emerging eco-habitus: the reconfiguration of high cultural capital practices among ethical consumers, *Journal of Consumer Culture* 14: 158–79.

Carrington D. (2021) Scientists convert used plastic bottles into vanilla flavouring – www.guardian.co.uk (15 June 2021).

Casemine (2019) R v Biffa Waste Services Ltd – www.casemine.com (last accessed 13 December 2022).

Casiella C. (2019) India's Mount Everest of trash – www.sciencealert.com (10 June 2019).

Castrica M., Tedesco D., Panseri S., Ferrazzi G., Ventura V., Frisio D. et al (2018) Pet food as the most concrete strategy for using food waste as feedstock within the European context: a feasibility study, *Sustainability* 10: 2035.

Cavendish R. (Glucksmann M.) (1982) *Women on the Line* (London: Routledge).

Chappells H., Shove E. (1999) The dustbin: a study of domestic waste, household practices and utility services, *International Planning Studies* 4: 267–80.

Chattopadhay S., Dutta A., Ray S. (2009) Municipal solid waste management in Kolkata, India, *Waste Management* 29(4): 1449–58.

Chi X., Wang M., Reuter M. (2014) E-waste collection channels and household recycling behaviours in Taizhou of China, *Journal of Cleaner Production* 80: 87–95.

Chilvers J., Burgess J. (2008) Power relations: the politics of risk and procedure in nuclear waste governance, *Environment and Planning A* 40: 1881–900.

China Highlights (2021) Chinese New Year food: top 7 lucky foods and symbolism – www.chinahighlights.com (28 January 2021).

Chiripanhura B., Wolf N. (2019) Long-term trends in UK employment: 1861–2018 – www.ons.gov.uk (15 December 2022).

Chrétien-Ichikawa S. (2015) Shanghai: a creative fashion system under construction, *China Perspectives* 2015/3: 33–41.

Clapp J. (2001) *Toxic Exports* (Ithaca: Cornell University Press).

Clapp J., Swanson L. (2009) Doing away with plastic shopping bags: international patterns of norm emergence and policy implementation, *Environmental Politics* 183: 315–32.

Clarke A. (1998) 'Window shopping at home: classifieds, catalogues and new consumer skills'. In Miller D. (ed) *Material cultures: why some things matter* (Chicago: University of Chicago Press), pp 73–99.

Clarke A. (2000) '"Mother swapping": the trafficking of nearly new children's wear'. In Jackson P., Lowe M., Miller D., Mort F. (eds) *Commercial cultures: economies, practices, spaces* (Oxford: Berg), pp 83–100.

Cloke P., May J., Williams A. (2017) The geographies of food banks in the meantime, *Progress in Human Geography* 41(6): 703–26.

Cockburn C., Furst-Dilic R. (eds 1994) *Bringing Technology Home* (Oxford: Oxford University Press).

Cockburn C., Ormrod S. (1993) *Gender and Technology in the Making* (London: Sage).

Cohen L. (2003) *A Consumer's Republic* (New York: Vintage).

Cooke P. (ed 1989) *Localities* (London: Unwin Hyman).

Cooper J. (2020) Rubbish from one of the first drive-thru KFCs to reopen in Wales left littering an entire area – www.walesonline.co.uk (27 May 2020).

Cooper T. (2010) Burying the 'refuse revolution': the rise of controlled tipping in Britain, 1920–1960, *Environment and Planning A* 42(5): 1033–48.

Coveya (2019) Recycling Glastonbury Festival – www.coveya.co.uk (last accessed 15 December 2022).

Coyle D. (2011) *The Economics of Enough* (Princeton NJ: Princeton University Press).

Crang M., Hughes A., Gregson N., Norris L., and Ahamed, F. (2013) Rethinking governance and value in commodity chains through global recycling networks, *Transactions Institute of British Geographers* 38: 12–24.

Creech L. (2020) Biffa loses appeal against waste export charges – resource. co (8 July 2020).

Creech L. (2021) China to ban all imports of solid waste from 2021 – resource.co (8 July 2020).

Crook L. (2019) BIG opens Copenhill power plant – www.dezeen.com (8 October 2019).

Cui C., Lui Y., Xia B., Jiang X., Skitmore M. (2020) Overview of public-private partnerships in the waste-to-energy incineration industry in China, *Energy Strategy Reviews* 32 https://doi.org/10.1016/j.esr.2020.100584.

Daily Post (2020) Calls for registration numbers to be stamped on fast food packaging after litter rise – www.dailypost.co.uk (5 June 2020).

Dant T. (2004) The driver-car, *Theory, Culture and Society* 21(4/5): 61–80.

Daunton M., Hilton M. (eds 2001) *The Politics of Consumption* (Oxford: Bloomsbury).

Davidson H. (2020) China to bring in law against food waste with fines for promoting over eating – www.guardian.co.uk (23 December 2020).

Davis D. (2006) 'Urban Chinese home owners as citizen consumers'. In Garon S., Maclachan P. (eds) *The ambivalent consumer* (Ithaca: Cornell University Press), pp 281–99.

Davis D. (ed 2000) *The Consumer Revolution in Urban China* (Berkeley CA: University of California Press).

De Kock R. (1987) The garbage scavengers: picking up the pieces, *Indicator South Africa* 4:51–55.

Defra (2004) Reducing litter caused by 'food on the go' – www.keepbrit aintidy.org (last accessed 13 December 2022).

Desrochers P. (2002) Industrial ecology and the rediscovery of inter-firm linkages: historical evidence and policy implications, *Industrial and Corporate Change* 11(5): 1031–57.

Desrochers P. (2008) Does the invisible hand have a green thumb? Incentives, linkages, and the creation of wealth out of industrial waste in Victorian England, *The Geographical Journal* 175 (1): 3–16.

Desrochers P., Lam K. (2008) Business as usual in the industrial age: (relatively) lean, green and eco-efficient? *The Electronic Journal of Sustainable Development* 1(1).

Dickens C. (2004 imprint) *Our Mutual Friend* (London: Penguin).

Di Nucci M., Brunnengräber A., Losada A. (2017) From the 'right to know' to the 'right to object' and 'decide': a comparative perspective on participation in siting procedures for high level radioactive waste repositories, *Progress in Nuclear Energy* 100: 316–25.

Dicken P. (2014, 7th edition) *Global Shift* (London: Sage).

Dietler M., Hayden B. (eds 2001) *Feasts* (Washington DC: Smithsonian Books).

Dodgson M. (2011) Exploring new combinations in innovation and entrepreneurship: social networks, Schumpeter and the case of Josiah Wedgwood (1730–1795), *Industrial and Corporate Change* 20: 1119–51.

Dolla T., Laisham B. (2021) Effects of energy from waste technologies on the risk profile of public-private partnership waste treatment projects of India, *Journal of Cleaner Production* 284: 124726.

Dorganova L. (2018) Discounting the future: a political technology, *Economic Sociology* 19(2): 4–9.

Dou Z., Toth J., Westendorf M. (2018) Food waste for livestock feeding: feeding, safety and sustainability implications, *Global Food Security – agriculture, policy, economics and environment* 17: 154–61.

Douglas M. (1966) *Purity and Danger* (London: Routledge & Kegan Paul).

Du Gay P., Hall S., James L., Madsen A., Mackay H., Negus K. (2013, 2nd edition) *Doing Cultural Studies* (London: Sage).

Dunajeva J., Kostka J. (2021) Racialized politics of garbage: waste management in urban Roma settlements in Eastern Europe, *Ethnic and Racial Studies* 45(9): 1738–1759.

Dwij (2019) The story of Asia's biggest textile recycling hub – www.dwijproducts.com (24 February 2019).

Easterlin R. (1974) 'Does economic growth improve the human lot?' In David P., Reder M. (eds) *Nations and Households in Economic Growth: essays in honour of Moses Abramowitz* (New York: Academic Press), pp 89–125.

Ebola Response Anthropology Platform: http://www.ebola-anthropology.net/ (last accessed 12 December 2022).

Ebrahimi H. (2012) Biffa owners eye quick fire sale after £1.2bn writedown – www.telegraph.co.uk (28 March 2012).

Economy E. (2004) *The River Runs Black* (Ithaca: Cornell University Press).

Eden S. (2017) Blurring the boundaries: prosumption, circularity and online sustainable consumption through Freecycle, *Journal of Consumer Culture* 17: 265–85.

Edmonds R. (ed 2000) *Managing the Chinese Environment* (Oxford: Oxford University Press).

Edwards F., Mercer D. (2007) Gleaning from gluttony: an Australian youth subculture confronts the ethics of waste, *Australian Geographer* 38(3): 279–96.

Edwards F., Mercer D. (2012) Food waste in Australia: the freegan response, *Sociological Review* 60: 174–91.

Ellen MacArthur Foundation (2012) Report Volume 1 – Towards the circular economy: economic and business rationale for an accelerated transition – https://ellenmacarthurfoundation.org/towards-the-circular-economy-vol-1-an-economic-and-business-rationale-for-an (last accessed 10 January 2023).

Ellen MacArthur Foundation (2013) Report Volume 2 – Towards the circular economy: opportunities for consumer goods – https://ellenmacarthurfoundation.org/towards-the-circular-economy-vol-2-opportunities-for-the-consumer-goods (last accessed 10 January 2023).

Ellen MacArthur Foundation (2014) Report Volume 3 – Towards the circular economy: accelerating the scale-up across global supply chains – https://ellenmacarthurfoundation.org/towards-the-circular-economy-vol-3-accelerating-the-scale-up-across-global (last accessed 10 January 2023).

Eminton S. (2013) Biffa completes £1bn recapitalisation – www.letsrecycle.co.uk (6 February 2013).

European Commission (nd) End of Life Vehicles – https://environment.ec.europa.eu/topics/waste-and-recycling/end-life-vehicles_en (last accessed 16 December 2022).

European Commission (nd) Landfill waste – https://environment.ec.europa.eu/topics/waste-and-recycling/landfill-waste_en (last accessed 16 December 2022).

Evans D. (2012) Beyond the throwaway society: ordinary domestic practice and a sociological approach to household food waste, *Sociological Review* 46: 41–56.

Evans D. (2014) *Food Waste* (London: Bloomsbury).

Evans D., Campbell H., Murcott A. (2012) A brief pre-history of food waste and the social sciences, *Sociological Review* 60: 5–26.

Expert Skip Hire (2019) Cleaning up the waste at Glastonbury – www.expertskiphire.co.uk (last accessed 13 December 2022).

Fahmi W. (2005) The impact of privatisation of solid waste management on the Zabaleen garbage collectors of Cairo, *Environment and Urbanisation* 17: 155–70.

Fantastic Fridges (nd) https://fantasticfridges.com/ (last accessed 10 December 2022).

Fareshare (nd) https://fareshare.org.uk/ (last accessed 8 December 2022).

Feedback (nd) Feeding People, Backing the Planet – https://feedbackglobal.org/ (last accessed 8 December 2022).

Ferrell J. (2006) *Empire of Scrounge* (New York: New York University Press).

Financial Times (2018) Asian millennials ditch stigma for buying second-hand – www.ft.com (27 September 2018).

Financial Times (2019) Second-hand market comes in from the cold in China – www.ft.com (30 July 2019).

Fitch Ratings (2021) Spotlight: China's Waste-to-Energy industry – https://www.fitchratings.com/research/corporate-finance/spotlight-china-waste-to-energy-industry-29-07-2021 (29 July 2021).

Ford M. (2015) *Rise of the Robots* (New York: Basic Books).

Ford M. (2021) *Rule of the Robots* (New York: Basic Books).

Frosch R., Gallopoulous N. (1989) Strategies for manufacturing, *Scientific American* 261: 144–52.

Furedy C. (1984) Survival strategies of the urban poor – scavenging and recuperation in Calcutta, *Geo Journal* 8(2): 129–34.

Gabrys J., Hawkins G., Michael M. (eds 2013) *Accumulation* (London: Routledge).

Galbraith J.K. (1958) *The Affluent Society* (London: Penguin).

Gandy M. (1999) The Paris sewers and the rationalisation of urban space, *Transactions Institute of British Geographers* 24(1): 23–44.

Garcier R. (2014) Disperse, confine or recycle? A geo-legal approach to the management and spatial circulations of low-level radioactive waste in France, *L'espace géographique* 43: 265–83.

Garrow D. (2012) Odd deposits and average practice: a critical history of the concept of structured deposition, *Archaeological Dialogues* 19: 85–115.

Gelber S. (1992) Free market metaphor: the historical dynamics of stamp collecting, *Comparative Studies in Society and History* 34: 742–69.

Georgescu-Roegen N. (1971) *The Entropy Law and the Economic Process* (Cambridge MA: Harvard University Press).

Gershuny J., Robinson J. (1988) Historical changes in the household division of labour, *Demography* 25: 537–52.

Gerth K. (2003) *China Made* (Cambridge MA: Harvard University Press).

Gerth K. (2010) *As China Goes, So Goes the World* (New York: Hill & Wang).

Gill K. (2007) Interlinked contracts and social power: patronage and exploitation in India's waste recovery market, *Journal of Development Studies* 43: 1448–74.

Gill K. (2009) *Of Poverty and Plastic* (Oxford: Oxford University Press).

Gille Z. (2007) *From the Cult of Waste to the Trash Heap of History: The Politics of Waste in Socialist and Postsocialist Hungary* (Bloomington: Indiana University Press).

Gille Z., Lepawsky J. (eds 2022) *The Routledge Handbook of Waste Studies* (London: Routledge).

Glastonbury Festival (1997–2022) Our waste policy – www.glastonburyfestival.co.uk (last accessed 12 December 2022).

Global Alliance for Incinerator Alternatives (2020) – www.no-burn.org (last accessed 15 December 2022).

Government Office for Science (2017) *From waste to resource productivity: Evidence and case studies* – www.assets.publishing.service.gov.uk (last accessed 12 December 2022).

Gowardhan M. (2020) How to celebrate Diwali – www.olivemagazine.com (last accessed 12 December 2022).

Graeber D. (2011) *Debt* (New York: Melville House).

Graham S., Thrift N. (2007) Out of order: understanding repair and maintenance, *Theory, Culture and Society* 24(3): 1–25.

Grand View Research (nd) Research Reports in Electronic & Electrical – https://www.grandviewresearch.com/industry/electronic-and-electrical (last accessed 10 December 2022).

Greenpeace (2021) Trashed – www.greenpeace.org.uk (last accessed 14 December 2022).

Gregory C. (1982) *Gifts and commodities* (London: Academic Press).

Gregson N. (2007) *Living with Things* (Oxford: Sean Kingston Publishing).

Gregson N., Brooks K., Crewe L. (2000) 'Narratives of consumption and the body in the space of the charity shop'. In Jackson P., Lowe M., Miller D., Mort F. (eds) *Commercial Cultures: Economies, Practices, Spaces* (Oxford: Berg), pp 101–22.

Gregson N., Crang M. (2015) From waste to resource: the trade in wastes and global recycling economies, *Annual Review of Environment and Resources* 40: 151–76.

Gregson N., Crang M. (2019) Made in China and the new world of secondary resource recovery, *Environment and Planning A: Economy and Space* 51(4): 1031–40.

Gregson N., Crang M., Botticello J., Calestani M., Krzywoszynska A. (2016) Doing the 'dirty' work of the green economy: resource recovery and migrant labour in the EU, *European Urban and Regional Studies* 23: 541–55.

Gregson N., Crang M., Ahamed F., Akter N., Ferdous R. (2010) Following things of rubbish value: end-of-life ships, 'chock-chocky' furniture and the Bangladeshi middle class consumer, *Geoforum* 41(6): 846–54.

Gregson N., Crang M., Ahamed F., Akter N., Ferdous R., Foisal S. et al (2012) Territorial agglomeration and industrial simbiosis: Sitakunda-Bhatiary, Bangladesh, as a secondary processing complex, *Economic Geography* 88(1): 37–58.

Gregson N., Crang M., Fuller S., Holmes H. (2015) Interrogating the circular economy: the moral economy of resource recovery in the EU, *Economy and Society* 44(2): 218–43.

Gregson N., Crang M., Laws J., Fleetwood T., Holmes H. (2013) Moving up the waste hierarchy: car boot sales, reuse exchange and the challenges of consumer culture to waste prevention, *Resources, Conservation and Recycling* 77: 97–107.

Gregson N., Crewe L. (1997) The bargain, the knowledge and the spectacle: making sense of consumption in the space of the car-boot sale, *Environment and Planning D: Society and Space* 15: 87–112.

Gregson N., Crewe L. (1998) Performance and possession: rethinking the act of purchase in the light of the car boot sale, *Journal of Material Culture* 2(2): 241–63.

Gregson N., Crewe L. (2003) *Second Hand Cultures* (Oxford: Berg)

Gregson N., Forman P. (2021) England's municipal waste regime: challenges and prospects, *The Geographical Journal* 187: 214–26.

Gregson N., Longstaff B., Crewe L. (1997) Excluded spaces of regulation: car-boot sales as an enterprise culture out of control?, *Environment and Planning A* 29:1717–37.

Gregson N., Metcalfe A., Crewe L. (2007a) Identity, mobility and the throwaway society, *Environment and Planning D: Society and Space* 25: 682–700.

Gregson N., Metcalfe A., Crewe L. (2007b) Moving things along: the conduits and practices of divestment in consumption, *Transactions Institute of British Geographers* 32: 187–200.

Gregson N., Metcalfe A., Crewe L. (2009) Practices of object maintenance and repair: how consumers attend to consumer objects within the home, *Journal of Consumer Culture* 9: 248–72.

Gregson N., Watkins H., Calestani M. (2013) Political markets: recycling, economization and marketization, *Economy and Society* 42(1): 1–25.

Gutierrez T., Webster G. (2012) Trash City: inside America's largest landfill site – www.cnn.com (last accessed 14 December 2022).

GWP Group (2020) Christmas packaging facts: the definitive list – www.gwp.co.uk (last accessed 12 December 2022).

Hannam K., Sheller M., Urry J. (2006) Editorial: mobilities, immobilities and moorings, *Mobilities* 1: 1–22.

Hansen K. (2000) *Salaula* (Chicago: Chicago University Press).

Hansen K., Le Zotte J. (2019) Changing second hand economies, *Business History* 61: 1–16.

Hanser A. (2010) Uncertainty and the problem of value: consumers, culture and inquality in China, *Journal of Consumer Culture* 10: 307–32.

Hardin G. (1968) The tragedy of the commons, *Science* 162 (3859): 1243048.

Harper G., Sommerville R., Kendrick E., Driscoll L., Slater P., Stolkin R. et al (2019) Recycling lithium-ion batteries from electric vehicles, *Nature* 575, 75–86.

Harper K., Sleger T., Filčák R. (2009) Environmental justice and Roma communities in Central and Eastern Europe, *Environmental Policy and Governance* 19: 251–68.

Harrison Evans P. (2016) *Shopping for Good: the Social Benefits of Charity Retail* (London: Demos).

Hartman H. (2010) Waste to energy processing plants – www.architectsjournal.co.uk (30 July 2010).

Hawkins G., Potter E., Race K. (2015) *Plastic Water* (Massachusetts: MIT Press).

Hayami Y, Dikshit A, Mishra A. (2006) Waste pickers and collectors in Delhi: poverty and environment in an urban informal sector, *The Journal of Development Studies* 42(1): 41–69.

Hebrok M., Heidenstrom N. (2019) Contextualising food waste prevention: decisive moments within everyday practices, *Journal of Cleaner Production* 210: 1435–48.

Hecht G. (2018) Interscalar vehicles for an African Anthopocene: on waste, temporality and violence, *Cultural Anthropology* 33(1).

Hellman K-U., Luedicke M. (2018) The throwaway society: a look in the back mirror, *Journal of Consumer Policy* 41: 83–57.

Henry L., Douhovnikoff V. (2008) Environmental issues in Russia, *Annual Review of Environment and Resources* 33: 437–60.

Herbert L. (2007) *A Centenary History of Waste and Waste Management in London and South East England* – Chartered Institute of Waste Management (www.ciwm.co.uk).

Herrman G., Soiffer S. (1984) For fun and profit: an analysis of the American garage sale, *H* 12: 397–421.

Hewitt D. (2007) *Getting Rich First* (London: Chatto & Windus).

Heyes C. (2006) Foucault goes to Weight Watchers, *Hypatia* 21: 126–49.

Hindustan Times (2018) In Panipat, the world's 'castoff capital', business hangs by a thread – www.hindustantimes.com (3 May 2018).

Hird M. (2012) Knowing waste: towards an inhuman epistemology, *Social Epistemology* 26(3–4): 453–69.

Hird M. (2021) *Canada's Waste Flows* (Montreal: McGill-Queens University Press).

Hochschild A. (1989) *The Second Shift* (New York: Viking).

Holmes H. (2018) New spaces, ordinary practices: circulating and sharing within diverse economic provisioning, *Geoforum* 88: 138–47.

Holmes H. (2019) Unpicking contemporary thrift: getting on and getting by in everyday life, *The Sociological Review* 67: 126–42.

Holt R., Popp A. (2016) Josiah Wedgwood, manufacturing and craft, *Journal of Design History* 29: 99–119.

Hoornweg D., Bhada-Tata P. (2012) *What a Waste: a global review of solid waste management* – www.openknowledge.worldbank.org (last accessed 12 December 2022).

Hu X., Wang C., Lim M., Koh S. (2020) Characteristics and community evolution patterns of the international scrap metal trade, *Journal of Cleaner Production* 243, 118576.

Hubbub to roll out 'on-the-go' recycling scheme to Scotland and Wales – www.edie.net (10 May 2019).

Huysman M. (1994) Waste picking as a survival strategy for women in Indian cities, *Environment and Urbanisation* 6(2): 155–74.

Independent (2015) Christmas food waste: over 4m festive dinners are thrown away – www.independent.co.uk (4 December 2015).

Inverardi-Ferri C. (2017) The enclosure of 'waste land': rethinking informality and dispossession, *Transactions of the Institute of British Geographers* 43(2): 230–44.

Investable Universe (2021) EQT in $5.3billion Take-private deal for Waste-to-energy leader Covanta – www.investableuniverse.com/2021/07/14/eqt-infrastructure-to-acquire-waste-to-energy-company-covanta/ (14 July 2021).

Jackman J. (2020) Glastonbury Festival's Carbon Footprint – www.theecoexperts.co.uk (last accessed 16 December 2022).

Jenkins D. (1990) Transatlantic trade in woollen cloth, 1850–1914: the role of shoddy – www.digitalcommons.unl.edu (last accessed 16 December 2022).

Jenkins D., Malin J. (1990) European competition in woollen cloth, 1870–1914: the role of shoddy, *Business History* 32(4): 66–86.

Jennings P. (2017) 'From waste plastics to marine power'. In Government Office for Science *From waste to resource productivity: Evidence and case studies* – www.assets.publishing.service.gov.uk (last accessed 16 December 2022), pp 59–62.

Jervis B. (2014) Middens, memory and the effect of waste: beyond symbolic meaning in archaeological deposits, an early medieval case study, *Archaeological Dialogues* 21: 175–96.

Johnson P. (1985) *Saving and Spending* (Oxford: Oxford University Press).

Joy A., Belk R., Wang J., Sherry J. (2020) Emotion and consumption: toward a new understanding of cultural collisions between Hong Kong and PRC luxury consumers, *Journal of Consumer Culture* 20: 578–97.

Kaza S., Yao L., Bhada-Tata P., van Woerden F. (2018) *What a Waste 2.0: a global snapshot of solid waste management to 2050* – www.openknowledge.worldbank.org (last accessed 16 December 2022).

Kellett B. (2014) A brief history of the wheelie bin – www.direct365.co.uk (30 June 2014).

Kent Online (2020) McDonald's drive-thru reopening causes outrage as litter left on roads – www.kentonline.co.uk (27 May 2020).

Kirklees Cousins West Yorkshire and Family History – https://kirkleescousins.co.uk/shoddy-and-mungo/# (last accessed 12 December 2022).

Kivell T. (2015) Evidence in hand: recent discoveries and the early evolution of human manual manipulation, *Philosophical Transactions of the Royal Society Series B* 370: 1682.

KKR enters UK's energy from waste sector – www.infrastructureinvestor.com (27 March 2020).

Kochetkova E., Pokid'Ko P. (2019) Soviet industrial production and waste disposal: a case study of pulp and paper plants on the Karelian isthmus, 1940s-1980s, *Scandinavian Economic History Review* 67(3): 269–82.

Kopytoff I. (1986) 'The cultural biography of things: commoditization as process'. In Appadurai A. (ed) The social life of things: commodities in cultural perspective (Cambridge: Cambridge University Press), pp 64–91.

Kristeva J. (1982) *Powers of Horror* (New York: Columbia University Press).

Kroen S. (2004) A political history of the consumer, *Historical Journal* 47(3): 709–36.

Kuppinger P. (2014) Crushed? Cairo's garbage collectors and neoliberal urban politics, *Journal of Urban Affairs* 36(S1 2): 621–33.

Laird K. (2016) Plaxx, a clean substitute for fossil-based heavy fuel oil? www.plasticstoday.co.uk (28 July 2016).

Lane R. (2011) The waste commons in an emerging resource recovery regime: contesting property and value in Melbourne's hard rubbish collections, *Geographical Research* 49: 395–407.

Langley J. (2021) Sale of Viridor assets to Biffa remains 'on track' – www.letsrecycle.co.uk (5 August 2021).

Langley P. (2021) Assets and assetization in financialized capitalism, *Review of International Political Economy* 28(2): 382–93.

Layard R. (2005) *Happiness* (London: Penguin).

Lee M. (2014) *Africa's World Trade* (London: Bloomsbury).

Lepawsky J. (2018) *Reassembling Rubbish* (Cambridge MA: MIT Press).

Lepawsky J., Mather C. (2011) From beginnings and endings to boundaries and edges: rethinking circulation and exchange through electronic waste, *Area* 43(3): 242–9.

Lepawsky J., McNabb C. (2010) Mapping international flows of electronic waste, *The Canadian Geographer/Le Géographie canadien* 54(2): 177–95.

Levison A. (2012) *The Fresh Kills Story: from the World's Largest Garbage Dump to a World-class Park* – documentary, Staten Island Borough President's Office – www.youtube.com (last accessed 13 December 2022).

Lewis S., Maslin A. (2015) Defining the Anthropocene, *Nature* 519 (7542): 171–80.

Liboiron M. (2019) Waste is not 'matter out of place' – www.discardstudies.com (9 September 2019) (last accessed 16 December 2022).

Liboiron M. (2022) 'Matter out of place'. In Gille Z., Lepawsky J. (eds) *The Routledge Handbook of Waste Studies* (London: Routledge), pp 31–40.

Loftus A., March H. (2016) Financialising desalination: rethinking the returns of big infrastructure, *International Journal of Urban and Regional Research* 40 (1): 46–61.

London B. (1932) 'Ending the depression through planned obsolescence', http://www.resol.com.br/textos/london_(1932)_ending_the_depression_through_planned_obsolescence.pdf (last accessed 10 January 2023).

Lora-Wainwright A. (2013) Introduction: dying for development: pollution, illness and the limits of citizen's agency in China, *The China Quarterly* 214: 243–54.

Lora-Wainwright A. (2021, revised edition) *Resigned Activism* (Massachusetts: MIT Press).

Mackie L. (2012) Don't let Biffa go to waste – www.guardian.co.uk (4 October 2012).

Makov T., Shepon A., Krones J., Gupta C., Chertow M. (2020) Social and environmental analysis of food waste abatement, *Nature Communications* 11: 1156.

Malinowski B. (1922) *Argonauts of the Western Pacific* (London: Routledge & Kegan Paul).

Marals H., Labushange J-P. (2019) If you want to prosper consider building roads: China's role in African infrastructure and capital projects – www.deloitte.com (22 March 2019).

Martin C. (2016) The sharing economy: a pathway to sustainability or a nightmarish form of neoliberal capitalism? *Ecological Economics* 121: 149–59.

MARTIN GmbH (nd) https://www.martingmbh.de/en/home.html (last accessed 13 December 2022).

Massey D. (1984) *Spatial Divisions of Labour* (London: Palgrave Macmillan).

Mathews G. (2011) *Ghetto at the Centre of the World* (Chicago: University of Chicago Press).

Mathews G., Ribeiro G., Vega C. (eds 2012) *Globalisation from Below* (London: Routledge).

Mathews G., Yang Y. (2012) How Africans pursue low end globalisation in Hong Kong and Mainland China, *Journal of Current Chinese Affairs* 41(2): 95–120.

Mauss M. (1925) Essai sur le don – The Gift: forms and functions of exchange in archaic societies – www.archive.org (last accessed 16 December 2022).

Mazke M. (1997) Precision grips, hand morpohology and tools, *American Journal of Physical Anthropology* 106: 91–110.

Mazke M. (2013) Tool making, hand morphology and fossil hominins, *Philosophical Transactions of the Royal Society Series B* 368(1630).

McCracken G. (1988) *Culture and Consumption* (Bloomington: Indiana University Press).

McKendrick N. (1960) Josiah Wedgwood: an C18th entrepreneur in salesmanship and marketing techniques, *Economic History Review* 12: 408–33.

McKendrick N., Brewer J., Plumb J. (1982) *The Birth of a Consumer Society* (Oxford: Oxford University Press).

McKinsey (2017) Automation, robotics, and the factory of the future – www.mckinsey.com (last accessed 16 December 2022).

McKinsey (2019) The state of fashion – www.mckinsey.com (last accessed 16 December 2022).

McMahon C. (2018) 'The Regulation and Development of the British Moneylending and Pawnbroking Markets, 1870–2016' – D Phil. Cambridge.

Medina M. (2007) *The World's Scavengers* (Lanham MD: AltaMira Press).

Melosi M. (2002) The Fresno sanitary landfill in a American cultural context, *The Public Historian* 24(3): 17–35.

Melosi M. (2008) *The Sanitary City* (Pittsburgh: University of Pittsburgh Press).

Melosi M. (2020) *Fresh Kills* (New York: Columbia University Press).

Merrell C. (2006) Guy Hands firm nets £1bn profit from waste disposal – www.thetimes.co.uk (18 July 2006).

Messenger B. (2017) Veolia to build 1.6m TPA waste to energy plant in Mexico City – www.waste-management-world.com (26 July 2017).

Metcalfe A., Riley M., Barr S., Tudor T., Robinson G., Guilbert S. (2012) Food waste bins: bridging infrastructures and practices, *The Sociological Review* 60(2): 135–55.

Meyer G. (1987) Waste recycling as a livelihood in the informal sector – the example of the refuse collectors of Cairo, *Applied Geography and Development* 30: 78–94.

Milanovic B. (2016) *Global Inequality* (Belknap: Harvard University Press).

Miller D. (1998) *A Theory of Shopping* (Cambridge: Polity).

Miller D. (2001) 'The birth of value'. In Jackson P., Lowe M., Miller D., Mort F. (eds) *Commercial cultures: economies, practices, spaces* (Oxford: Berg), pp 77–84.

Miller D. (ed 1995) *Acknowledging Consumption* (London: Routledge).

Miller D. (ed 2002) *Car Cultures* (Oxford: Berg).

Miller D., Jackson P., Thrift N., Holbrook B., Rowlands M. (1998) *Shopping, Place and Identity* (London: Routledge).

Ministry of Housing and Urban Affairs (India) (nd) – *Transforming Urban Landscapes of India*. https://pdfroom.com/books/transforming-urban-landscapes-of-india/zW5n19Ww2Nq (last accessed 5 February 2023).

Minter A. (2013) *Junkyard Planet* (London: Bloomsbury).

Minter A. (2018) With new clothes as cheap as used ones, Panipat's recycling industry goes out of fashion – www.economictimes.com (16 January 2018).

Minter A. (2020) China finally makes its peace with 'foreign garbage' – www.bloomberg.com (29 July 2020).

Mohn J., Szidat S., Fellner J., Rechberger H., Quartier R., Buchmann B. et al (2008) Determination of biogenic and fossil CO_2 emitted by waste incineration based on $^{14}CO_2$ and mass balances, *Bioresource Technology* 99: 6471–79.

Moore S. (2008) The politics of garbage in Oaxala, Mexico, *Society and Natural Resources* 21(7): 597–610.

Morley D. (1986) *Family Television* (London: Routledge).

Morley N., Bartlett C., McGill I. (2009) *Maximising reuse and recycling of UK clothing and textiles*: a report to the Department for Environment, Food and Rural Affairs, Oakdene Hollins Ltd. https://www.oakdenehollins.com/reports/2009/12/1/maximising-reuse-and-recycling-of-uk-clothing-and-textiles-report (1 December 2009).

Myers F. (ed 2001) *The Empire of Things* (Santa Fe: School of American Research Press).

Narishkin A., Cameron S., Barranco V., Hauwanga K. (2021) How $300m Carnival cruise ships are demolished in Turkey – www.businessinsider.com (29 July 2021).

National Museum of American History – Food: transforming the American taste – www.americanhistory.si.edu (last accessed 16 December 2022).

NDTV Food (2020) Diwali 2020 food: delicious Diwali food menu for a memorable lunch and dinner party – www.food.ndtv.com (last accessed 12 December 2022).

NDTV Food (2020) Eid 2020: full Eid party menu with recipes to make your Eid al Fitr celebrations delicious – www.food.ndtv.com (last accessed 16 December 2022).

Needham S., Spence T. (1997) Refuse and the formation of middens, *Antiquity* 271: 77–90.

Nelson M., Rademacher M., Paek H-J. (2007) Downshifting consumer = unshifting citizen? An examination of a local Freecycle community, *Annals of the American Association of Political and Social Science* 611: 141–56.

Neuwirth R. (2011) *Stealth of Nations* (New York: Anchor Books).

New Economics Foundation – https://neweconomics.org/ (last accessed 12 December 2022).

Nixon J., Dey P., Ghosh S. (2017) Energy recovery from waste in India: an evidence-based analysis, *Sustainable Energy Technologies and Assessments* 21 (June): 23–32.

Norris L. (2012) Economies of moral fibre? Recycling charity clothing into emergency blankets, *Journal of Material Culture* 17: 389–404.

NS Energy (nd) Shenzhen East waste-to-energy plant – www.nsenergybusiness.com (last accessed 16 December 2022).

Nuclear Decommissioning Authority (2019) Integrated Waste Management: Radioactive Waste Strategy.

O'Brien M. (2013) Consumers, waste and the 'throwaway society' thesis: some observations on the evidence, *International Journal of Applied Sociology* 3: 19–27.

O'Connor K. (2011) *Lycra* (London: Routledge).

O'Hare P. (2019) 'The landfill has always borne fruit': precarity, formalisation and dispossession among Uruguay's waste pickers, *Dialectical Anthropology* 43(1): 31–44.

OLIO – https://olioex.com (last accessed 8 December 2022).

O'Neill K. (2019) *Waste* (Cambridge: Polity).

Ongka's Big Moka (1974) Dir. Charlie Nairn (Granada: Disappearing World).

Oteng-Arabia M. (2012) When necessity begets ingenuity: e-waste scavenging as a livelihood strategy in Accra, Ghana, *African Studies Quarterly* 13(1&2): 1–21.

Packard V. (1960) *The Waste Makers* (New York: Ig Publishing).

Pahl R. (1984) *Divisions of Labour* (Oxford: Basil Blackwell).

Park A. (2021) Waste management companies grow more attractive to private equity firms – www.koreatimes.co.kr (17 December 2021).

Parry J. (1986) The Gift, the Indian gift and the 'Indian gift', *Man* 21: 453–73.

Pellow D. (2008) 'The global waste trade and envrionmental justice struggles'. In Gallagher K. (ed) *Handbook on Trade and the Environment* (Cheltenham: Edward Elgar).

Persaud D., Lepawsky J., Liboiron M. (2019) Viscous objects: the uneven resistance of repair, *Techniques & Culture* 72: https://journals.openedition.org/tc/12372.

Piketty T. (2014) *Capital in the Twenty First Century* (Cambridge MA: Harvard University Press).

Pitts M., Pattie, C., Dorling D. (2007) Christmas feasting and social class: Christmas feasting and everyday consumption, *Food, Culture and Society* 10: 407–24.

Pour N., Webley P., Cook P. (2018) Potential for using municipal solid waste as a resource for bioenergy with carbon capture and storage, *International Journal of Greenhouse Gas Control* 68: 1–15.

Press and Journal (2016) Litter explosion after opening of drive-through McDonalds in Inverness – www.pressandjournal.co.uk (13 October 2016).

Provencher J., Bond A., Avery-Gomm S. Borelle S., Rebolledo E., Hammer S. et al (2017) Quantifying ingested debris in marine megafauna: a review and recommendations for standardisation, *Analytical Methods* 9: 1454–69.

Pryke M., Allen J. (2019) Financialising urban water infrastructure: extracting local value, distributing value globally, *Urban Studies* 56: 1326–1346.

Quinault C. (2015) Biffa eyes more acquisitions as it returns to growth – www.letsrecycle.co.uk (22 April 2015).

Ramsay N. (2009) Taking-place: refracted enchantment and the habitual spaces of the tourist souvenir, *Social and Cultural Geography* 10: 197–217.

Rathje W. (1984) The garbage decade, *The American Behavioural Scientist* 28: 9–29.

Rathje W., Murphy C. (1992) *Rubbish: the archaeology of garbage* (Tucson: University of Arizona Press).

Reinert H., Reinert E. (2006) 'Creative destruction in Economics: Nietzsche, Sombart, Schumpeter'. In Backhaus J., Drechsler W. (eds) *Friedrich Nietzsche (1844–1900): The European heritage in Economics and the social sciences 3* (Springer: New York), pp 55–85.

Rejh R. (2021) Waste management in India – an overview, *United International Journal for Research and Technology* 2(7).

Restart – https://therestartproject.org/ (last accessed 10 December 2022).

Ricci D. (2020) *A Political Science Manifesto for the Age of Populism* (Cambridge: Cambridge University Press).

Rivoli P. (2005) *Travels of a T shirt in the Global Economy* (New Jersey: John Wiley and Sons).

Roberts A., Zulfiqar G. (2019) Social reproduction, finance and the gendered dimensions of pawnbroking, *Capital and Class* 43: 581–97.

Roosevelt A., Housley R., Da Silveira M., Maranca S., Johnson R. et al (1991) Eighth Millenium pottery from a pre-historic shell midden in the Brazilian Amazon, *Science* 254 (5038): 1621–24.

Rootes C. (2006) Explaining the outcomes of campaigns against waste incineration in England: community, ecology, political opportunities and policy contexts, *Research In Urban Policy* 10: 179–198.

Rootes C., Leonard L. (2010) *Environmental Movements and Waste Infrastructure* (London: Routledge).

Rosa E., Cirelli C. (2018) Scavenging: between precariousness, marginality and access to the city, *Environment and Planning A – Economy and Space* 50: 1407–24.

Rowlingson K. (2013) Short of cash, rent and food – Britons in dire financial straits – www.theconversation.com (25 July 2013).

Ryan A., Rothman R. (2022) Engineering chemistry to meet COP26 targets, *Nature Reviews Chemistry* 6: 1–3.

Sadler J., Wallace S. (2021) Microbial synthesis of vanillin from waste polyethylene terephthalate, *Green Chemistry* 2021 Jun 10;23(13): 4665-4672. doi: 10.1039/d1gc00931a. eCollection 2021 Jul 5.

Sanderson H. (2022) *Volt Rush* (London: One World).

Schlosser E. (2002) *Fast Food Nation* (London: Penguin).

Schulz Y. (2015) Towards a new waste regime? Critical reflections on China's shifting market for high-tech discards, *China Perspectives* (2015/3): 43–50.

Schumpeter J. (1942) *Capitalism, Socialism, Democracy* (London: Routledge).

Schwartz-Cohen R. (1985) *More Work for Mother* (New York: Basic).

Segal T. (2021) Enron Scandal: the Fall of a Wall Street Darling – www.investopedia.com/updates/enron-scandal-summary/ (26 November 2021).

Shapiro-Bengsten S. (2020) Is China building more incinerators than it needs? – www.chinadialogue.net (12 August 2020).

Sharholy M., Ahmad K., Mahmood G., Trivedi R. et al (2008) Municipal solid waste management in Indian cities – a review, *Waste Management* 28(2): 459–67.

Shell H. (2014) Shoddy heap: a material history between waste and manufacture, *History and Technology* 30: 374–94.

Shell H. (2020) *Shoddy* (Chicago: Chicago University Press).

Sheller M., Urry J. (2007) The new mobilities paradigm, *Environment and Planning A* 28: 207–26.

Shi H., Chertow M., Song Y. (2010) Developing country experience with eco-industrial parks: a case study of the Tianjin Economic-Technological Development Area in China, *Journal of Cleaner Production* 18(3): 191–99.

Shove E. (2003) *Comfort, Cleanliness and Convenience* (Oxford: Berg).

Shove E., Pantzar M., Watson M. (2012) *The Dynamics of Social Practice* (London: Sage).

Silverstone R. (1994) *Television and Everyday Life* (London: Routledge).

Silverstone R., Hirsch E. (eds 1992) *Consuming Technologies* (London: Routledge).

Simon D., Brooks A. (2012) Unravelling the relationships between used-clothing imports and the decline of African clothing industries, *Development and Change* 43(6): 1265–90.

Skobla D., Filčák R. (2020) Mundane populism: politics, practices and discourses of Roma oppression in rural Slovakia, *Sociologica Ruralis* 60: 773–89.

Slavin T. (2016) An incinerator with a view: Copenhagen waste plant gets ski slope and picnic area – www.guardian.co.uk (26 October 2016).

Song J., Song D., Zhang X., Sun Y. (2013) Risk identification for PPP waste-to-energy incineration projects in China, *Energy Policy* 61(C): 953–62.

Song G., Zhang H., Duan H., Xu M. (2018) Packaging waste from food delivery in China's mega cities, *Resources, Conservation and Recycling* 130 (March 2018): 226–27.

Southerton D. (2006) Analysing the temporal organisation of daily life: social constraints, practices and their allocation, *Sociology* 40: 435–54.

Southerton D. (2013) Habits, routines amd temporalities of consumption: from individual behaviours to the reproduction of everyday practices, *Time & Society* 22: 335–55.

Southerton D., Shove E. (2000) Defrosting the freezer: from novelty to convenience. A story of normalization, *Journal of Material Culture* 5: 301–19.

Spigal L. (1992) *Make Room for TV* (Chicago: University of Chicago Press).

Stallybrass P. (1998) 'Marx's coat'. In Spyer P. (ed) *Border Fetishisms* (London: Routledge), pp 183–207.

Steffen W., Crutzen P., McNeill J. (2007) The Anthropocene: are humans now overwhelming the great forces of nature? *Ambio* 36: 614–21.

Steffen W., Grinevald T., Crutzen P., McNeill J. (2011) The Anthropocene: conceptual and historical perspectives, *Philosophical Transactions of the Royal Society Series A* 369: 342–67.

Strasser S. (1999) *Waste and Want* (New York: Metropolitan Books, Henry Holt).

Strathern M. (1988) *The Gender of the Gift* (Berkeley: University of California Press).

Stuart T. (nd) https://www.tristramstuart.co.uk/ (last accessed 8 December 2022).

Sullivan R. (2020) How the world's largest garbage dump evolved into a green oasis – www.nytimes.com (14 August 2020).

Talyan V., Dahiya R., Sreekrishnan T. (2008) State of municipal waste management in Delhi, the capital of India, *Waste Management* 28(7): 1276–87.

Tarasuk V., Dachner N., Loopstra R. (2014) Food banks, welfare and food insecurity in Canada, *British Food Journal* 116(9): 1405–17.

Taylor D. (2021) UK waste firm fined £1.5m for exporting household waste – www.guardian.co.uk (30 July 2021).

Tebbutt M. (1984) *Making Ends Meet* (Leicester: Leicester University Press).

Tevere D. (1994) Dump scavenging in Gabarone, Botswana: anachronism or refugee occupation of the poor, *Geografiska Annaler* 76(B) 1: 21–32.

The Economist (2017) Panipat, the global centre for recycling textiles is failing – www.economist.com (9 September 2017).

The National Pawnbrokers Association of the UK – https://www.thenpa.com/ (last accessed 11 December 2022).

The Sun (2017) The Morning After – www.thesun.co.uk (26 June 2017).

Thomas J. (2012) Some deposits are more structured than others, *Archaeological Dialogues* 19: 124–27.

Thomas N. (1991) *Entangled Objects* (Cambridge MA: Harvard University Press).

Thompson E.P. (1963) *The Making of the English Working Class* (London: Victor Gollancz).

Thompson M. (1979) *Rubbish Theory* (Oxford: Oxford University Press).

Thorpe N. (2016) The place where cast-off clothes end up – www.bbc.co.uk (13 June 2016).

Thrift N. (2004) Driving in the city, *Theory, Culture and Society* 21(4/5): 41–59.

Times of India (2018) Diwali food menu for lunch and dinner party – www.timesofindia.indiatimes.com (2 November 2018).

Tocheri M., Orr C., Jacofsky M., Marzke M. (2008) The evolutionary history of the hominin hand since the last common ancestor of *Pan* and *Homo*, *Journal of Anatomy* 212: 544–62.

Tolvik Consulting (2021) UK Energy from Waste Statistics – 2020 (www.tolvik.com) (last accessed 16 December 2022).

Tong X., Tao D. (2015) The rise and fall of a 'waste city' in the construction of an 'urban circular economic system': the changing landscape of waste in Beijing, *Resources, Conservation and Recycling* 107 (February): 10–17.

Trentmann F. (2006) 'Knowing consumers – histories, identities, practices: an introduction'. In Trentmann F. (ed) *The Making of the Consumer: knowledge, power and identity in the modern world* (Oxford: Berg), pp 1–27.

Trentmann F. (2016) *Empire of Things* (London: Allan Lane).

Tsang E. (2014) *The new middle class in China* (Basingstoke: Palgrave Macmillan).

UK Radioactive Waste Inventory – https://ukinventory.nda.gov.uk/ (last accessed 12 December 2022).

UKRI (2022) Commercialisation of plastic waste derived fuel for generating electricity – gtr.ukri.org (last accessed 12 December 2022).

United Nations, Department of Economic and Social Affairs (nd) Sustainable Development – https://sdgs.un.org/ (last accessed 8 December 2022).

United Nations Economic Commission (nd) Managing Methane for a better Climate – epa.gov/ghgemi – https://unece.org/unece-and-sdgs/managing-methane-better-climate (last accessed 8 December 2022).

United States Environmental Protection Agency - https://www.epa.gov/ghgemissions/overview-greenhouse-gases (last accessed 8 December 2022).

Unravel (2012) Dir. Meghna Gupta (Soul Rebel Films) – www.soulrebelfilms.com.

Urry J. (2007) *Mobilities* (Cambridge: Polity).

Van der Horst H., Pascucci S., Bol W. (2014) The 'dark side' of food banks: exploring emotional responses of food bank receivers in the Netherlands, *British Food Journal* 116(9): 1506–20.

Van der Zwan N. (2014) Making sense of financialisation, *Socio-economic Review* 12(1): 99–129.

Veblen T. (1994 edition) *The Theory of the Leisure Class* (London: Penguin).

Veolia (nd) Solving the disposable cup conundrum – www.livingcircular.veolia.com (last accessed 12 December 2022).

Veolia (2019) Annual financial report 2018 – www.veolia.com (last accessed 14 December 2022).

Veolia (2021a) Half-yearly annual financial report 2021 – www.veolia.com (last accessed 16 December 2022).

Veolia (2021b) Integrated report – The benchmark company for the ecological transformation – www.veolia.com (last accessed 14 December 2022).

Veolia (2021c) Presentation – Creating the world champion of the ecological market – www.veolia.com (last accessed 16 December 2022).

Verschoor A., van Gelderen A., Hofstra U. (2021) Fate of recycled tyre granulate used on artificial turf, *Environmental Sciences Europe* 33.

Visual Storytelling Team in London (2021) The EV revolution: how green is your vehicle? – www.ft.com (5 October 2021).

Von Fossen A. (2016) Flags of convenience and global capitalism, *International Critical Thought* 6(3): 359–77.

Wagner T. (2014) Reducing single use plastic shopping bags in the USA, *Waste Management* 70: 3–12.

Walsh D. (2002) Urban residential refuse compostion and generation rates for the twentieth century, *Environment, Science and Technology* 36: 4936–42.

Warde A. (1999) Convenience food: space and timing, *British Food Journal* 101: 518–27.

Warde A. (2005) Consumption and theories of practice, *Journal of Consumer Culture* 5: 131–53.

Warde A., Cheng S-L., Olsen W., Southerton D. (2007) Changes in the practice of eating: a comparative analysis of time use, *Acta Sociologica* 50: 363–85.

Washburn S. (1960) Tools and human evolution, *Scientific American* 203(3): 62–75.

Waters C., Zalsiewicz J., Summerhays C., Barnosky A., Poirier C., Galuszka A. et al (2016) The Anthropocene is functionally and stratigraphically distinct from the Holocene, *Science* 351 (6269).

Watson M., Meah A. (2012) Food waste and safety: negotiating conflicting social anxieties into the practices of domestic provisioning, *Sociological Review* 60: 102–20.

Wei R., Zimmermann W. (2017) Microbial enzymes for the recycling of recalcitrant petroleum-based plastics: how far are we? *Microbial Biotechnology* 10(6): 1308–22.

Weiner A. (1992) *Inalienable Possessions* (Berkeley CA: University of California Press).

Which? (2021) Appliances that last (January), p 25.

White M. (2018) Plastic: recycling the unrecyclable? – www.themanufacturer.com (5 November 2018).

Widdall C. (2018) Shoddy and mungo – kirklesscousins.co.uk (26 March 2018).

Wiens K. (2018) European policy makers plot a path past planned obsolescence – ifixit.com (26 September 2018).

Wilkie R. (2019) Why are energy from waste schemes so troublesome? – www.constructionmanagermagazine.com (14 March 2019).

Williams C. (2020) Viridor 'to sell waste and recycling subsidiary' – www.mrw.co.uk (2 October 2020).

Wilson D., Rodic L., Scheinberg A., Velis C., Alabaster G. et al (2012) Comparative analysis of solid waste management in 20 cities, *Waste Management and Research* 30(3): 237–54.

Wilson D., Velis C., Cheeseman C. (2006) Role of informal sector recycling in waste management in developing countries, *Habitat International* 30(4): 797–808.

Wolfe A. (2008) Nylon: a revolution in textiles – www.sciencehistory.org (2 October 2008).

World Economic Forum (2016) The new plastics economy: rethinking the future of plastic.

WRAP (2020a) Food waste and COVID-19 – survey 3 – www.wrap.org (last accessed 16 December 2022).

WRAP (2020b) Food surplus and waste in the UK – www.wrap.org (last accessed 16 December 2022).

Wright A. (2018) How a California landfill became a Landmark – www.atlasobscura.com (last accessed 16 December 2022).

Yang X. (2019) China pushes garbage sorting, a costly but worthwhile endeavour – www.news.ctgn.com (6 September 2019).

Yo L. (2014) *Consumption in China* (Cambridge: Polity).

Zhang D., Huang G., Xu Y., Gong Q. et al (2015) Waste-to-energy in China: key challenges and opportunities, *Energy* 8(12): 14182–196.

Zhang W. (2017) No cultural revolution? Continuity and change in consumption patterns in contemporary China, *Journal of Consumer Culture* 17: 639–58.

Zhao J. (2014) *The Chinese Fashion Industry: An Ethnographic Approach* (London: Bloomsbury).

Zhao X., Jiang G-W., Li A., Wang L. (2016) Economic analysis of waste to energy in China, *Waste Management* 48: 604–18.

Zheng L., Song J., Li C., Gao Y., Geng P., Qu B. et al (2014) Preferential policies promote MSW to energy in China: current status and prospects, *Renewable and Sustainable Energy Reviews* 36: 135–48.

Zhu Q., Lowe E., Wei Y., Barnes D. (2007) Industrial symbiosis in China: a case study of the Guitang Group, *Journal of Industrial Ecology* 11(1): 31–42.

Zimmerman F. (1951) *World Resources and Industries* (New York: Harper and Row).

Zuboff S. (2019) *The Age of Surveillance Capitalism* (London: Profile Books).

Zuto (nd) Drive-thru density – www.zuto.com (last accessed 16 December 2022).

Index

References to endnotes show both the page number and the note number (185n2).

A

abjection 41–3
activist approach 7–8, 11–12, 14, 152–3
  see also environmental campaigners
Africa 200–1n57
Ambikapur (India) 150
anaerobic digestion 144–5
animals
  companion animals 14–15, 185n2
  ingestion of plastic 32, 185n3
  livestock 14
Anthropocene 33, 185n7
Appadurai, A. 78
ash residue 95–6
assets/assetization 134, 162, 197n2, 197n3
auto culture *see* drive-thru outlets

B

Babcock & Wilcox 155
baby clothes/baby things 85–6
Bangladesh 129–30
Batley (UK) 112
Bhatiary (Bangladesh) 129–30
Biffa 161, 201–2n74, 202n76
bins 39–40, 94–8, 188–9n6
biogas 144–5, 150–1, 178
biosecurity 184n30
biotechnologies 123–4, 126–7, 144–5
blending of recovered materials 112–13, 131–2
body boundaries 41–2
boilers 153, 155–6
brands 70
'British Restaurants' programme 169, 202n1
bulk purchasing 13, 67–8

C

calorific value of waste 147–8, 151, 155, 157, 201n61
campaign groups *see* activist approach; environmental campaigners

capital
  finance capital 160–3
  fixed capital 153–6
  operating capital 156–60
capitalism, discard in theories of 52–4
capitalization 134, 148
  see also capital;
    commodification; financialization
car boot sales 94
car culture, eating, drinking and 39, 40–1, 43, 186n21
car industry 57
car recovery/scrappage 119–20
car repair 69, 72
car tyres 127–8
carbon budgets/allowances 173
carbon capture 175
carbon emissions 173, 174–5
carbon offsetting 174–5
caste 37, 186n15
cats 15
charitable giving 87–8, 100–1, 114
charity bags 87, 88
charity food sector 10
chemical industry 175–6, 177
chemical technologies 123–6
Chhattisgarh (India) 150
China
  energy-from-waste 154
  festival 43–4
  food waste policy 182n21
  global investments by 160, 200–1n57
  significance of consumption in 58–9
  waste management regimes 146–8
Christmas discard 51
Christmas food 44, 45
circular economy 32, 88, 102, 124, 128, 129, 146, 196–7n72
cities 3–4, 39–40, 96, 138–40, 150–2, 158
climate emergency 179
  see also environmental concerns

227

climate politics 173, 174–5
climate science 135, 143
clothing
  baby clothes 85–6
  obsolescence 66–8
  practices related to 19–20
  repair 73
  valuation 50, 101
clothing discard viii–ix, 50, 85–6, 101
clothing/textiles recycling
  global shifts 111–14
  sorting process 114–18
CO_2 *see* carbon emissions
collect-curate model 175–6, 177–8
collectable items 21
collection-disposal regime
  emergence of 138–40
  in India 148–50
  turn to landfill 140–2
  US landfills 142–3, 198n15, 198n16, 198–9n18
collection workers 37, 81–2, 114, 149–50, 152, 178
  see also scavenging
colonialism 107–8
commodification
  of waste 27–8
  see also recommodification of discard
commodities 78
commodity markets 121, 123
commodity traders 107, 131
companion animals 14–15, 185n2
compost/composting 145, 150–1, 178, 199–200n47
construction, of energy-from-waste plants 156
construction waste 1–2
  see also house renovation
consumer behaviour 56–7
consumer carbon budgets 173
consumer demand xii, 56–7, 58–9, 169–74
consumer-derived waste *see* municipal waste
consumer discard
  product obsolescence 59–60
    clothing 66–8
    home entertainment 63–6
    white goods 60–3
  relationship with economies 54–5
consumer durables
  obsolescence 60–3
  repair 69–70, 72–3
consumer goods, intermittent use of 76–7
consumer-heavy economies xiii
  consumer discard in 59–68
  emergence of 55–9
consumerism 17–18
consumers
  ethical 166–7
  re-imagining 173–4

consumption
  and activist approach to waste 8, 11
  author's research on viii
  benefits 170
  challenge of reconfiguring 166–9
  digitally mediated 172
  economic significance 54–9
  experience-oriented 171, 173
  and food surplus 10–11
  implications of used goods exchange 101–3
  and municipal waste xiii
  overconsumption 10–11, 87, 90–1
  and policy approach to waste 4–5
  practices of 19–20, 71–5
  relationship with energy-from-waste 158
  relationship with recycling 165
  relationship with waste 4–5
  social context of 167
  understandings of 4–5, 17–20
  and valuation 20–3
controlled tipping 141–2
convenience
  of giving conduits 87, 90
  and household bins 98–9
  questioning reliance on 168–9
  of replacement versus repair 72–3
  and societal change 12–13, 62
convenience food 13, 15
  see also fast food
convenience shopping 13–14
cooking practices 11, 13, 45, 168
Copenhill plant 200n55
corporate interests, in energy-from-waste 153–6, 158–60
corporeal reality 41–2
COVID-19 pandemic 5, 57–8, 167–8, 183n24
Cream 115, 116
creative destruction 53, 188n1
credit services *see* pawnbrokers
curation 175–6, 177

D

data 3–4, 172, 177, 180–1n1
death, discard following 48
decarbonization 174–6
delivery workers 168–9, 178
demand
  consumer demand xii, 56–7, 58–9, 169–74
  for waste 133, 158, 165
demolition 190n28
demolition waste 1–2
dependence effect 56, 57
Diderot effect 62, 189–90n20
digital technologies 65–6, 171–3
  see also online platforms
dioxin emissions 152
dirt 35, 36, 186n11
'dirty' work 37–8

INDEX

discard
  concept of 21, 31–2
  food *see* food waste/discard
  from festivals 45–7
  household bin studies 188–9n6
  recommodification *see* recommodification of discard
  reconfiguring consumption and 166–9
  relationship with waste 31, 51, 106, 129, 197n74
  rights to 148
  and temporality 47–50
  in theories of economic activity 52–4
  value regimes 80–1
    cascade of goods through 99–101
    giving 81, 83–91
    sale/credit 81–2, 91–4
    waste/binning 82–3, 94–8
  see also waste
discard management
  reorienting industry to 175–6
  rescaling waste management 177–8
discard studies 31–2, 185n1
  see also waste studies
discarding
  eating and drinking 'on the go' 38–41, 43
  as embodied activity 34
  and festival 43–7
  as human activity 31–2, 50
  and life transitions 47–50
  surges in 51
disposability 67–8, 192–3n27
disposable goods 22, 39, 184n44, 187n24, 187n25
disposal technologies *see* biotechnologies; chemical technologies; incineration; landfill
dissolution x
Diwali food 44
dogs 15
domestic technologies 60–3, 72–3
domestic work 12, 74, 75
domesticated farm animals 14
Douglas, Mary 35
drink *see* food and drink
drive-thru outlets 40, 186n21
dumps and dumping 24, 107–8, 139–40, 150
  see also landfill
DuPont 67
durable value 21
durables *see* consumer durables

E

eating *see* food and drink
eBay 192n9
ecological civilization 146, 148
ecological imaginary xi
ecological principles 32

economic activity
  discard in theories of 52–4
  residue/waste from 24–5
economic approach 24–6
economic development 108–9
economic dimension of waste xi, xiii
economic growth 4, 5, 170–1
economic organization, discard in theories of 52–4
economic policy 55–7
economic restructuring 53, 188n3
economies
  discard in consumer-heavy economies 59–68
  repair contingent on 70
  significance of consumption 54–9
  see also circular economy
Eid food 44
electricity 157
electronic discard 171–2
emerging markets 59
  see also global South
emissions *see* GHGs
employment
  consumer-related 56–7
  in textile recycling 112
  women's 12, 61, 62
  see also waste workers
End-of-life Vehicles Directive (EU) 119, 195–6n51
energy-from-waste
  biotechnologies 144–5
  and climate politics 174–5
  finance capital 160–3
  fixed capital 153–6
  incineration reconfigured as 135, 136–7, 144
  in India 150
  operating capital 156–60
  opposition to 152–3
  relationship with recycling 145, 201n62
  rise of 151–2
energy-from-waste plants 153–6, 200n55
England
  textile recycling in 111–13
  waste management regimes in 139–42
enough/sufficient 8, 182n13
environmental campaigners 27, 97, 106–7, 120–1, 152–3
  see also activist approach
environmental concerns 179
  and economic growth 170
  and experience-oriented consumption 173
  waste management policy 174–5
  see also climate science; policy interventions
environmental justice 25, 107
environmental service companies 158–60
environmental stewardship 88, 89
ethical consumption 166–7

229

Europe
  energy-from-waste sector 154, 158–9
  see also United Kingdom
European Union (EU)
  changing waste regimes 136–7, 143–4, 145
  End-of-life Vehicles Directive 119, 195–6n51
  politics of waste reduction 6
  recycling regulation 128
evolutionary biology 33
exchange see gift exchange
experience-oriented consumption 171, 173

F

Fareshare 183n22
farm animals 14
farming (urban) 178
fashion 63, 66–8
fast fashion 67, 68
fast food 39–41, 43
feasting 44–6
festivals 43–7
'final mile' logistics 168–9, 178
finance capital 134, 160–3
financial value of discard (sale/credit) 81–2, 91–4
financialization 134, 197n2
  of waste xiii, 28, 161–3
fixed capital 153–6
food and drink
  feasting 44–6
  and mobility 38–41, 43, 186–7n23
  policy interventions 57–8, 169, 202n1
  as troublesome 42–3
food manufacture and processing 62
food poverty 169, 182n17
food preparation 11, 13, 45, 168
food retail sector 9, 10, 13, 62
food storage 61–3, 72, 73
food waste/discard 9–15, 169, 182n16, 183n24, 185n2
  collection of 178
  and festival 45–6
  and mobility 38–41, 43, 186–7n23
  waste-to-resource 144–5, 147
football shirts 66
free market economics 57
Freecycle 85, 88–90
Fresh Kills landfill/Park 142–3, 198n15
Fresno landfill site 143, 198n16
fridges/fridge-freezers 61–3, 72, 73
frugality 166–9
  see also waste prevention/waste saving
furnace grates 154–5

G

Galbraith, J.K. 56, 57
garbage archaeology 188–9n6
garden waste 6–7
gate fees 140

gendered division of labour 12, 60–1, 74, 75
GHGs (greenhouse gases) 6, 174
  CO_2 173, 174–5
  methane 27, 135, 136, 143, 185n52, 202n3
gift exchange 46, 83–4, 85–91
gift wrapping 46
giving 81, 83–91
Glastonbury festival 46–7
global flows of discard 105–7
  China 109–10
  toxic colonialism 107–8
  used clothing 115, 116–17
global North
  collection-disposal regime 138–42
  flow of discard from 105–7, 108
  future of recycling in 130–1
  recommodification in 118–20, 121–2, 123
  waste-to-resource regime 142–6
  see also Europe; United States
global recycling networks xi
global shifts in textile recycling 111–14
global South
  economic development 108–9
  and recommodification 105–10, 129–30
  and toxic colonialism 107–8
  waste management regimes 146–52
globalization xi, 57, 173
globalization from below 105–7
Government Office for Science (UK) xiv
grate technologies 154–5
Great Depression 55–6
greenhouse gases see GHGs
growth see economic growth
Gumtree 192n9

H

'hand-me-down-around' economy 86
hierarchy of waste 136, 176–7
home entertainment goods 63–6, 72, 190n23
homeworking 167–8, 178
horses, valuation of 79–80
hosiery 67–8
hospitality sector 9–10, 57–8
house moving 48, 98
house renovation 50
house repair 69, 190n28
household bins 94–8, 188–9n6
household recycling 121–3, 147–8, 149–50
housing market 54
human hand, capacities of 33–4

I

ICT see information and communication technology
identities 18
IKEA 102
incineration
  in China 146–7
  cost of 140

230

and economics of waste 27
historical view 139–40
in India 151
opposition to 152
policy interventions 174–5
and politics of waste 6
reconfigured as energy-from-waste 135, 136–7, 144
see also energy-from-waste
India 113–14, 148–51
industrial discard 129
industrial ecology 32, 128, 146, 196–7n72, 199n25
industrial restructuring 53, 188n3
informal sector 37, 81–2, 114, 148, 149–50, 152
information and communication technology (ICT) 65–6, 72, 171–3
see also online platforms
innovation 175
and obsolescence 53–4, 59–60
clothing 66–8
home entertainment 63–6
white goods 60–3
in recovery/recycling 110–11, 112
biological and chemical technologies 123–7
mechanical technologies 119–23
investors, in energy-from-waste 163

J
Jenkins, D. 195n27

K
Kershaw, John 197n74
Keynes, John Maynard 55
knowledge xv–xvii
Kristeva, Julia 41
kula ring 84, 85–6

L
labour conditions, in textile recycling 112
landfill
in China 146
and economics of waste 26–7
in India 148–9, 150
move away from 135, 136, 143–4, 174
policy interventions 6, 27, 136, 143–4, 174, 202n3
and politics of waste reduction 6
turn to 140–2
US examples 142–3, 198n15, 198n16, 198–9n18
landfill restoration 141, 198n15
'last-mile' logistics 168–9, 178
Law, Benjamin 111–12, 132
leasing 171
Liboiron, Max 35
life transitions, and discard 47–50

limited-production runs 21
litter 35–6, 39–40, 187n27
livestock 14
living arrangements (changing) 48–9
London, Bernard 60

M
Malinowski, B. 44, 84
manufactured limited-production runs 21
manufactured obsolescence 61, 66–8, 69
marginal utility 56
Martin Gmbh 154–5
Marxist theory 52, 78
materiality 71–2
materials recovery facilities (MRFs) 122–3
measurement, politics of 5–7
mechanical and biological treatment (MBT) 144–5
mechanical technologies 112, 119–23
memory work 78–9
messages, from research findings xv, xvii
metal recovery 119–20
methane 27, 135, 136, 143, 185n52, 202n3
methodological nationalism ix
Mexico City 151
microplastics 128, 185n3
middens 141
mobility ix–x
eating, drinking and 38–41, 43
moka exchanges 84
moving house 48, 98
MRFs (materials recovery facilities) 122–3
mungo 194n22
municipal waste
evolution of 16–17
focus on xiii–xiv
residual waste within 136
scale of 2, 3–4
in waste studies xii–xiii
municipal waste regimes 94–8
music festivals 46–7
mutability x

N
negative externalities 24, 27
'net zero' 174, 175
New Suez 159
nuclear industry 2, 181n4, 181n6

O
obsolescence 22, 53–4, 59–60
clothing 66–8
home entertainment 63–6
white goods 60–3
Olio 182n16
online platforms, for exchange 82, 85, 88–90, 93–4, 192n9
operating capital 156–60
overconsumption 10–11, 87, 90–1

P

Packard, Vance 60, 189n16
Panipat (India) 113–14
pawnbrokers 81, 91
Pemex 107
pets 14–15, 185n2
pigs 14
planned obsolescence 63, 65–6, 68, 172
plastic bags and packaging 2, 182n10
plastic discard, ingestion of 32, 185n3
plastic pollution 124, 128
plastic waste, recovery/recycling of 124–7, 196n58, 201n62
Plaxx 125–6
policy domain
  and the academy xiv–xvii
  approach to waste in 3–7
policy interventions
  discard management 176
  electronic discard 172
  incineration 174–5
  landfill 6, 27, 135, 136, 143–4, 174, 202n3
  public health 138–9
  recovery/recycling regulation and 119, 128, 129, 130, 131, 172, 195–6n51
  repairability 69–70, 172
  travel and carbon emissions 173
pollution 7, 24, 25
  food and drink as source of 42
  plastics 124, 128
potlatch 84
PPP (public–private partnerships) 149, 158, 162, 163
practices of consumption 19–20, 71–5
Preloved 192n9
private equity sector 160–2
production processes, discard from 52–3
public health 96, 138–40
public–private partnerships (PPP) 149, 158, 162, 163
Puente Hills landfill site 143

Q

quality
  and calorific value of waste 147–8, 151, 155, 157, 201n61
  of recovered materials 112–14, 122–3, 132
  of used clothing 115

R

radioactive waste 2, 181n4, 181n6
'rag and bone' collectors 81–2, 114
recommodification of discard 104–5
  biological and chemical technologies 123–7, 144–5
  China 109–10
  environmental challenges of 127–8
  future of recycling 128–32
  global flows xi, 105–7

in global North 118–20, 121–2, 123
mechanical technologies 119–23
textile recycling 111–18
toxic colonialism 107–8
recovery, and residual waste management 136–7
recovery/recycling
  biochemical technologies 123–7
  cars 119–20
  in China 109–10, 147–8
  concept of 104–5
  of digital technology 172
  environmental challenges 127–8
  future of 128–32
  Glastonbury festival 47
  household recycling collections 121–3, 149–50
  in India 149–50
  innovation 110–11, 112
  mechanical technologies 119–23
  plastics 124–7, 196n58, 201n62
  policy/regulatory interventions 119, 128, 129, 130, 131, 172, 195–6n51
  relationship with consumption 165
  relationship with energy-from-waste 145, 201n62
  ships 120–1, 129–30
  textile recycling 111–18
recycling networks (global) xi, 105–7
refurbished goods 70
regulation see policy interventions
religion, and festival 43
rent-based models of consumption 171
repair 8, 68, 69–75, 89, 168, 172
replacement 53, 60
  see also obsolescence
resale markets 91–4
residual waste 133–4, 136–7
  see also energy-from-waste
residue trading 107–8
retail sector
  food retail 9, 10, 13, 62
  sites 53
reuse 8, 102
  textile reuse markets 115, 116–17
  see also used/secondhand goods
reverse imperialism 108
Ricci, D. 188n1
rights, to discard/waste 25, 148, 156–7
Roma 37, 186n17
rubbish see discard
rubbish value 21, 22, 23, 82–3, 94–8

S

sale/credit 81–2, 91–4
salvage 104
sanitary disposal 138–40
sanitary landfill, rise of 141–2
scarcity, and value 21

scavenging 25, 37, 104, 193n1
scrap market 70, 119–20, 129–30
  see also recovery/recycling
secondhand goods *see* used/secondhand goods
self/subjectivity 41–3, 49
selling/credit 81–2, 91–4
service sector 57–8
sharing economy 76, 88, 171
ship recycling 120–1, 129–30
shipping, use of Plaxx 125
shoddification 131–2
shoddy 111–13, 132, 194n22, 194n24, 195n27
single-use items *see* disposable goods
singularization 90
smart fridge-freezers 63
smart TVs 64–5
social context of consumption 167
social economy 81, 83–91, 191n7
social science, and policy environment xv–xvii
social stratification 37–8
societal celebration 43–7
societal change 11–14, 75
  and convenience 12–13, 62
Sony Walkman 22, 184n43
sorting/segregation
  in China 147–8
  in collect-curate model 177
  of household recycling 122–3, 147, 149–50
  in India 149–50
  informal sector 148, 149–50
  of used clothing 115–17
  see also quality
source segregation 147–8
special purchase vehicles (SPVs) 162
steel production 129–30
stockings 67
storage xi, 26, 185n51
student graduation, discard following 48–9
Suez (later New Suez) 159
sufficient/enough 8, 182n13
supermarkets 9, 10, 13, 62
surplus
  concept of 8
  as consumer goods category 76–7
  and food waste 9–10
  value regimes and discard of 80–1
    cascade of goods 99–101
    giving 81, 83–91
    sale/credit 81–2, 91–4
    waste/binning 82–3, 94–8
surplus food 9–10, 169
Swachh Bharat Mission 149, 150, 177
swap events 85

T

technological innovation
  and obsolescence 59–60
    clothing 66–8
    home entertainment 63–6
    white goods 60–3
  in recovery/recycling
    biological and chemical technologies 123–7
    mechanical technologies 112, 119–23
    textile recycling 112
technologies
  biotechnologies 123–4, 126–7, 144–5
  in energy-from-waste sector 144–5, 153–6
  ICT/digital 65–6, 72, 171–3
    online platforms 82, 85, 88–90, 93–4, 192n9
  and recovery sector in global North 118–19
  and source segregation in China 147
television consumption research 190n22
televisions 63–5
textile recycling
  global shifts 111–14
  sorting process 114–18
thing agency 20
Thompson, Michael 21, 23
thrift
  and food practices 11, 13, 45
  potential for 166–9
  see also waste prevention/waste saving
throwaway society 55, 189n7
Timarpur (India) 151
time/temporality
  and cascade of goods 100, 101
  and changing consumption 168
  and convenience 12–13, 62
  and discard 47–50
  and repair 72–3
tipping 141–2, 150
  see also landfill
toxic colonialism 107–8
Trafigura/Probo Koala case 107–8
transient value 21–2
transnational companies 154–6, 158–60
trash *see* discard
travel, emissions burden of 173
Trobriand Islanders 44, 84
tyres 127–8

U

uncontrolled tipping 141
United Kingdom
  energy-from-waste sector 159, 161, 201n69
  textile recycling 111–13
  waste management regimes 139–42
United States
  collection-disposal regime 142–3
  landfill sites 142–3, 198n15, 198n16, 198–9n18
  urban areas 3–4, 39–40, 96, 138–40, 150–2, 158
urban cultivation 178
Urbaser 159–60
use-by dates 13

used/secondhand goods
  clothing 85–6, 101, 114–18
  giving 81, 83–91
  global flows of 106, 109
  implications for consumption 101–3
  repairability 69, 70
  for sale/credit 81–2, 91–4
  valuation 89, 91, 99, 100–1, 103

V

valuation 79–80
  and clothing discard 50
  and consumption 20–3
  and financialization 134, 197n4
  and used goods 89, 91, 99, 100–1, 103
  in value regimes 99, 100–1
  see also sorting/segregation
value 79–80
  and repair 69
  of residual waste 156–8
value categories 21–2, 23
value regimes 77–9
value regimes of discard 80–1
  cascade of goods through 99–101
  giving 81, 83–91
  intersection of 99
  sale/credit 81–2, 91–4
  waste/binning 82–3, 94–8
Van der Zwan, N. 197n2
vanillin 126–7
Veolia 158–9
Viridor 161

W

washing machines 60–1, 72–3
waste
  binning as value regime 82–3, 94–8
  commodification 27–8
  demand for 133, 158, 165
  financialization xiii, 28, 161–3
  increasing 3–4, 133, 165
  inevitability of 164
  policy see policy domain;
    policy interventions
  relationship with discard 31, 51, 106, 129, 197n74
  residue from economic activity 24–5
  scale of waste problem 1–4
  scholarship on vii, xi–xii
  see also discard
waste activist approach 7–8, 11–12, 14
  see also environmental campaigners
waste collectors 37, 81–2, 114, 149–50, 152, 178
  see also scavenging
waste commons 25, 148, 149
waste data 3–4, 180–1n1
waste generation, politics of reducing 5–7
waste hierarchy 136, 176–7

waste management
  and climate politics 174–5
  food, drink and mobility 39–40, 187n25
  Glastonbury festival 47
  innovation 175
  policy see policy interventions
  politics of 6
  re-scaling 177–8
  reconfiguration of 175–6
  and surges of discarding 51
  see also energy-from-waste;
    incineration; landfill
waste management regimes
  collection-disposal regime 138–42
  concept of 137–8
  global North 138–46
  global South 146–52
  municipal waste regimes 94–8
  waste-to-resource regime 82, 94–8, 99, 101
    in global North 142–6
    in global South 146–52
waste prevention/waste saving
  activist focus on 7–8, 11, 14
  potential for thrift/frugality 166–9
  relationship with consumption 101–3
  thrifty food practices 11, 13, 45
waste studies vii, xi–xii
  see also discard studies
waste-to-resource regime 82, 94–8, 99, 101
  in global North 142–6
  in global South 146–52
  see also energy-from-waste
waste-to-wealth, in India 149
waste types
  construction and demolition 1–2
  electronic 171–2
  food see food waste/discard
  garden 6–7
  harmful 164
  plastic see plastic waste
  in policy domain 3
  radioactive 2, 181n4, 181n6
  residual waste 133–4, 136–7
  sorting see sorting/segregation
  see also municipal waste
waste workers 37–8, 148
  collectors and pickers 37, 81–2, 114, 149–50, 152, 178
water services 158–9, 162
weight of waste 6–7, 119
wheelie bins 96–7
white goods 60–3, 72–3
women's employment 12, 61, 62
woollen industry 111–14
work
  domestic work 12, 74, 75
  textile recycling 112

women in labour market 12, 61, 62
see also waste workers
working arrangements
  homeworking 167–8, 178
  transition and discard 49

World Bank 3–4
wrapping 46

Z

Zero Waste argument 133, 197n1

www.ingramcontent.com/pod-product-compliance
Lightning Source LLC
Chambersburg PA
CBHW070920030426
42336CB00014BA/2469